CIRIA C547

Scoping the assessment of sediment plumes from dredging

S A John
S L Challinor
M Simpson
T N Burt
J Spearman

sharing knowledge ■ building best practice

6 Storey's Gate, Westminster, London SW1P 3AU
TELEPHONE 020 7222 8891 FAX 020 7222 1708
EMAIL enquiries@ciria.org.uk
WEBSITE www.ciria.org.uk

Summary

CIRIA Research Project 600 reviews the current state of knowledge on sediment plumes arising from dredging different bed materials, in a variety of environments using several dredging techniques.

The need for this project was identified through CIRIA consultation with the dredging industry, regulators and other stakeholders. Information on sediment plumes is extensive but disparate, and needs to be brought together in documents that can be used by developers and regulators alike to provide a structured approach to the assessment of the effects of sediment plumes arising from dredging.

This scoping report, therefore, provides a review of research work undertaken on the plumes arising from marine aggregate and capital and maintenance dredging and examines current knowledge on the environmental impacts. In doing this, it considers the current understanding of dredging plumes, the existing mechanisms for knowledge transfer and identifies gaps in knowledge. It then recommends further work required to fill these gaps and to develop a comprehensive framework for assessing the effects of sediment plumes, by defining the necessary components of such a framework.

Scoping the assessment of sediment plumes from dredging

John, S A, Challinor, S L, Simpson, M, Burt, T N and Spearman, J

Construction Industry Research and Information Association

CIRIA Publication C547   © CIRIA 2000   ISBN 0 86017 547 2

Keywords	
Sediment plumes, aggregate dredging, capital dredging, maintenance dredging	
Reader interest	**Classification**
Dredging industry, regulators and consultants concerned with aggregate, capital and maintenance dredging.	AVAILABILITY Unrestricted CONTENT Guidance document STATUS Committee-guided USER Dredging industry and regulators

Published by CIRIA, 6 Storey's Gate, Westminster, London SW1P 3AU. All rights reserved. No part of this publication may be reproduced or transmitted in any form or by any means, including photocopying and recording, without the written permission of the copyright-holder, application for which should be addressed to the publisher. Such written permission must also be obtained before any part of this publication is stored in a retrieval system of any nature.

Acknowledgements

This report was produced as a result of CIRIA Research Project 600 "Scoping the assessment of sediment plumes arising from dredging". The project was carried out on behalf of CIRIA by a consortium led by Posford Duvivier Environment and included HR Wallingford and the Centre for Environment, Fisheries and Aquiculture Science (CEFAS)

CIRIA's research managers were Daniel J Leggett and Stuart Meakins. Following CIRIA's established practice, the research project was guided by a steering group, comprising of:

Russell Arthurton (chairman)	Independent consultant
Graham Boyes	MAFF
John Land	Dredging Research Ltd
Paul Leonard	MAFF
Chris Vivian/Tom Matthewson	CEFAS
Sue Rees	English Nature
Frank Rimell	FDC
Ian Selby	BMAPA
Tom Simpson	DETR Minerals and Waste Planning Division
Ad Stolk	Ministry of Transport, Public Works and Water Management North Sea Directorate, the Netherlands
Gerard Van Raalte	Hydronamic BV

The following were corresponding members:

Richard Allen	Harwich Haven Authority
Douglas Clarke	Department of the Army Waterways Experimental Station, USA
Anders Jensen	Danish Hydraulic Institute
Gerard Loman	Boskalis Westminster Dredging BV
Jim Rodger	Mouchel Asia Ltd

CIRIA is grateful to the following organisations that provided financial support to the project:

DETR Minerals and Waste Planning Division

British Marine Aggregate Producers Association

English Nature

Federation of Dredging Contractors

MAFF

Ministry of Transport, Public Works and Water Management North Sea Directorate, the Netherlands.

Contents

Abbreviations 9

1 Introduction 11
    1.1 Project background 11
    1.2 The report... 12
    1.3 Dredging in the UK and worldwide 17
    1.4 Dredging techniques 19
    1.5 Sediment plumes... 22

2 Legislation, regulation and stakeholders 27
    2.1 Legislation and regulation.. 27
    2.2 Stakeholders.. 30
    2.3 Knowledge transfer between stakeholders... 33

3 Sources of sediment plumes 35
    3.1 Introduction... 35
    3.2 Effect of dredging technique on source strength... 36
    3.3 Measurement of sediment losses 48
    3.4 Recent field studies 52

4 Plume processes and modelling 55
    4.1 Summary of research 55
    4.2 Processes affecting dynamic plumes.. 57
    4.3 Processes affecting passive plumes.. 59
    4.4 Modelling techniques... 63

5 Effects on the marine environment 67
    5.1 Introduction... 67
    5.2 Effects on water quality 70
    5.3 Effects on marine ecology 77
    5.4 Effects on fish 84
    5.5 Effects on shellfish.. 90
    5.6 Other environmental effects 93
    5.7 Cumulative effects 97

6 Mitigating environmental effects 103
    6.1 Choice and operation of dredging plant.. 103
    6.2 Environmental windows 108

7 Towards and assessment framework 111
    7.1 The requirements of an assessment framework 111
    7.2 Gaps in current knowledge 115
    7.3 Improving modelling techniques 118

8 Recommendations – the way forward 119
    8.1 Developing a comprehensive assessment framework 119
    8.2 Adoption of a draft assessment framework: research priorities 119
    8.3 Addressing knowledge gaps 120
    8.4 Developing best practice guidance for assessing environmental effects... 124
    8.5 Recommendations for knowledge transfer... 127

References 129

Appendices 143

A1 **Legislation and regulation** 143
    A1.1   Marine aggregate dredging 143
    A1.2   Capital and maintenance dredging 145

A2 **Sediment plume stakeholders** 149
    A2.1   The dredging industry 149
    A2.2   Regulators 150
    A2.3   Other interested parties 151
    A2.4   Knowledge transfer 152

A3 **Field measurements of losses from dredging operations** 155
    A3.1   Trailing suction hopper dredging measurements 155
    A3.2   Cutter suction measurements 159
    A3.3   Grab dredging measurements 160
    A3.4   Backhoe dredging measurements 162
    A3.5   Bucket ladder dredging measurements 163
    A3.6   Measurements from other dredgers 164

A4 **Recent field studies** 167
    A4.1   Plume measurement system of the USACE Dredging Research Program 167
    A4.2   Channel deepening to Londonderry, Lough Foyle, Northern Ireland 168
    A4.3   Agitation dredging at Sheerness, eastern England 170
    A4.4   Construction of pipeline trench, southern England 171
    A4.5   Maintenance dredging of the River Tees, north-eastern England 171
    A4.6   Aggregate dredging operations in the UK 172
    A4.7   Dredging activities in Hong Kong 173
    A4.8   Minipod measurements at Area 107 and Race Bank 174
    A4.9   Montoring associated with the Øresund Link, Denmark and Sweden 174

A5 **Modelling techniques** 177
    A5.1   Dredging process modelling 177
    A5.2   Dynamic plume 177
    A5.3   Mud flow/concentrated suspension modelling 179
    A5.4   Passive plume modelling 179
    A5.5   Water quality models 181

A6 **A comparison of predictive approaches of sediment plume dispersion** 183
    A6.1   Introduction 183
    A6.2   Desk analysis 183
    A6.3   Analytical advection/diffusion mpodelling 184
    A6.4   Process modelling – SEDPLUME 185
    A6.5   Methodology 186
    A6.6   Results for uniform flow conditions – offshore from Harwich Harbour. 186
    A6.7   Effect of Lagrangian residuals 187
    A6.8   Non-unidirectional flow 188
    A6.9   Other considerations 188
    A6.10 Conclusions of comparisons of different approaches to plume dispersion prediction 188

List of figures

1.1	A user's guide to the process and the scoping report	10
1.2	Limits of the technical scope of the report	15
1.3	Dynamic plume phase (due to overflowing)	23
1.4	Passive plume phase	25
2.1	Primary interactions between stakeholders in the UK aggregate dredging industry and capital and maintenance dredging industry	32
3.1	Trailing suction hopper dredger used for aggregate dredging in the UK	38
3.2	Cutter suction dredger	40
3.3	Grab dredging	41
3.4	Bed leveller dredging	44
3.5	Plumes formed by hydrodynamic dredging	45
3.6	Hydrodynamic dredging in the Bristol Channel	46
3.7	Cutter suction dredger *Orion* dredging through chalk and clay – south coast of England	48
3.8	Changes in spill and production by cutter suction dredger *Castor* 1996–1997	54
5.1	A guide to the effects on the marine environment	68
7.1	Indicative components of an assessment framework	112

List of boxes

1.1	Definition of relevant terminology	13
1.2	Effects on the marine environment	17
1.3	Dredging quantities in the UK	18
2.1	US Army Corps of Engineers	34
3.1	Use of the sediment flux method during the Øresund Link project	53
5.1	Benthic ecology – r and K strategists	79
5.2	Effects of sediment plume settlement (ERM, 1999)	81
5.3	Effects of suspended sediments on the behaviour of adult herring and cod	88
5.4	Effect of sediment plumes on berried hen crabs at Area 107, Race Bank	92
5.5	CEFAS Research Project – cumulative environmental impacts of marine aggregate extraction	99
A2.1	The UK Marine SACs project	153
A5.1	The GAUSSIAN model	181

List of tables

1.1	World totals for dredger type	19
3.1	Summary of the characteristics of different dredging projects	36
3.2	Example values for the mass of sediment re-suspended and lost during dredging (mg/l and kg/s)	50
3.3	Indicative values for the mass of sediment re-suspended per m^3 of dredged material	51
3.4	Summary of the spillage from cutter suction dredging	53
3.5	Summary of the spillage from dipper and backhoe dredging	53
4.1	Physical processes affecting sediment plume dispersion	55
5.1	Environmental effects of heavy metals	73
5.2	Environmental effects of hydrocarbon and OCI compounds	75
5.3	Coincidence of dredging with benthic communities	78
5.4	Particle size distributions of overspill and reject material	94
5.5	Screened load quantities	94
5.6	The US CEA process	98
6.1	Choice and operation of dredging plant	103
6.2	USACE environmental windows survey regarding sediment plumes	109
A3.1	Measurements of turbidity around a trailing suction hopper dredger	156
A3.2	Typical sediment disturbances around a trailing hopper dredge	157
A3.3	Trailing suction hopper measurements at Lough Foyle	158
A3.4	Summary of UK spillway measurement losses	159
A3.5	Measurements of turbidity around a cutter suction dredger	159
A3.6	Cutter suction dredging measurements at Nieuwe Binnenhaven	160
A3.7	Dutch measurements of turbidity around a grab dredger	161
A3.8	US measurements of turbidity around a grab dredger	162
A3.9	Measurements of turbidity around a bucket ladder dredger	163
A3.10	Measurements of turbidity around a disc cutter dredger	164
A3.11	Measurements of turbidity around a pneuma pump dredger	165
A3.12	Measurements of turbidity around a wormwheel suction dredger	166
A6.1	Model parameters	186

Abbreviations

BMAPA	British Marine Aggregate Producer's Association (UK)
CCW	Countryside Council for Wales (UK)
CEC	Crown Estate Commissioners (UK)
CEDA	Central Dredging Association
CEFAS	Centre for Environment, Fisheries and Aquaculture Science (UK)
DETR	Department of the Environment, Transport and the Regions (UK)
DFI	District Fisheries Inspectorate (UK)
DOE	former Department of the Environment, now under the DETR (UK)
DOENI	Department of the Environment for Northern Ireland (UK)
EIA	environmental impact assessment
IADC	International Association of Dredging Contractors
IAPH	International Association of Ports and Harbours
LMOB	light material overboard
MAFF	Ministry of Agriculture, Fisheries and Food
MAGIS	Marine Sand and Gravel Information Service
MMS	Minerals Management Service, US Department of the Interior (USA)
PIANC	Permanent International Association of Navigation Congresses
QPA	Quarry Products Association (UK)
SAMS	Scottish Association of Marine Science (UK)
SAC	Scottish Association of Marine Science (UK)
SFC	Sea Fisheries Committee (UK)
SNH	Scottish Natural Heritage (UK
SODD	Scottish Office Development Department (UK)
SPA	Special Protection Area
USACE	US Army Corps of Engineers (USA)
USEPA	US Environmental Protection Agency (USA)
VBKO	Vereniging van Waterbouwers in Bagger-, Kust-en Oeverwerken, the Netherlands
WODA	World Organisation of Dredging Associations

A comprehensive glossary of dredging terms is provided in the Permanent International Association of Navigation Congress's (PIANC) *Glossary of dredging terms* (in preparation, 2000).

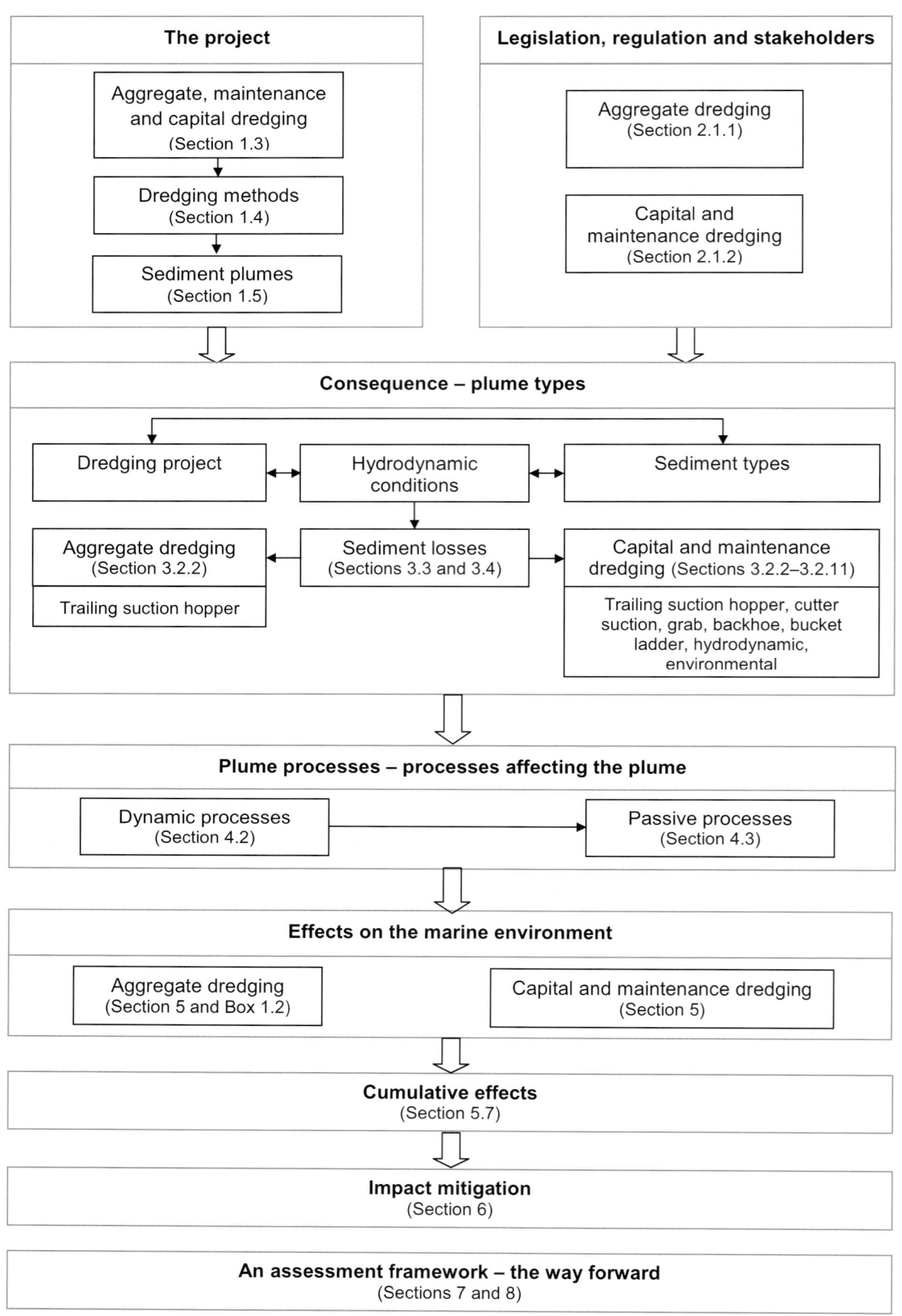

Figure 1.1 *A user's guide to the process and the scoping report*

1 Introduction

1.1 PROJECT BACKGROUND

Sediment plumes arising from dredging operations are perceived by many stakeholders in the marine and coastal environment to cause environmental effects that require investigation and monitoring. The dredging may be undertaken for any one of a number of reasons, such as the extraction of marine aggregates for construction, to gain material for beach recharge, land reclamation and infrastructural works, to deepen waterways as part of capital works or for the maintenance of berths and navigation channels. Less often it is carried out to "clean-up" contaminated water bodies. The dredging industry and government agencies have therefore undertaken or commissioned research studies into the potential impacts of sediment plumes. Research has also been commissioned into the various sediment plumes that arise from dredging different materials using a variety of dredging techniques. Considerable progress has been made on the mathematical modelling of sediment plumes and on the field measurement of plumes using acoustic and optical techniques, supported by on-site sampling and calibration. However, verified predictive techniques and knowledge of the environmental effects of sediment plumes are lacking.

Under current UK legislation, all dredging licence applications and some maintenance dredging operations are subject to environmental impact assessments (EIA). All EIAs include extensive consultation with stakeholders, including the regulators, conservation agencies and fishery organisations. Most assessments include a review of the potential impacts arising from the generation of sediment plumes. Specifically, assessments tend to investigate the environmental effects associated with the sediment plume while it is in the water column and its subsequent settlement to the seabed.

During these phases, sediment plumes have the potential to affect various environmental parameters. Water quality can be affected, for example, by increasing the biological oxygen demand of the water column or releasing previously bound-up contaminants. Marine ecology, both on the seabed and in the water column, can be affected due to physical processes such as shading or smothering. Shellfisheries can also potentially be impacted by sediment plumes as a result of effects on their filter-feeding efficiency. Similarly, impacts on fisheries can arise due to the physical alteration of spawning and nursery areas and direct physiological impacts on juvenile and adult fish.

With this background, CIRIA Research Project (RP) 600 was established to review the current state of knowledge on the nature of sediment plumes arising during dredging operations and their effects. The need for the project was identified through consultation with the dredging industry, regulators and other stakeholders, in response to the commonly held view that current knowledge on sediment plumes is too disparate. It needed to be brought together for concurrent use by both developers and regulators to improve the assessment of the plume effects of dredging. It is also clear that there is a need for a structured approach to the assessment of the effects of sediment plumes arising from dredging.

RP600 considers these issues by addressing the following objectives:

- to review research and relevant case studies (in the UK and elsewhere) of sediment plumes arising from marine aggregate and capital and maintenance dredging
- to examine current knowledge on the environmental impacts arising from sediment plumes
- to review the relevance of legislation relating to sediment plumes
- to identify gaps in current knowledge and research initiatives to address them
- to produce a comprehensive report on current understanding
- to recommend further work required to produce guidance, review modelling techniques and develop a comprehensive framework for the assessment of the environmental effects associated with dredging plumes.

In achieving these objectives, this scoping report draws together the extensive knowledge on sediment plumes and focuses on the requirements for better data collection and monitoring. Building from this, the next phase of the work will be able to move forward to the provision of comprehensive information on plume effects that can be used by developers, regulators and consultees alike, and a structured approach to the assessment of impacts. Meeting these objectives will ease the assessment process, help to achieve consensus regarding impacts and impact significance, and lead to cost and time savings in EIA and monitoring.

1.2 THE REPORT

1.2.1 Technical scope

The research team was tasked with producing a **scoping report** on the assessment of sediment plumes arising from dredging operations for both marine aggregate and capital and maintenance dredging in marine, coastal and estuarine areas. Plumes arising from the dredging of inland waterways are beyond the scope of this report. The research was objective-based and required the provision of information on the various parts of the dredging industry, different dredging techniques, legislation and regulation, sediment plume dynamics and modelling, and environmental effects. Relevant definitions for terms used throughout the report are included in Box 1.1.

Sediment plumes may also arise from placement of the dredged material for beneficial use or more conventional disposal. Much research has been undertaken on the plume processes and environmental effects generated during disposal operations, some of which is relevant to this project. International conventions such as the London Convention 1972 and the Oslo Paris Convention encourage the beneficial use or recycling of dredged material. However, the assessment of sediment plumes arising from dredged material placement is outside of the scope of RP600, which focuses on the sediment plumes arising directly from dredging operations in open water and specifically excludes disposal operations.

Similarly, the report does not consider the re-suspension after sediment has settled from a plume onto the seabed. However, the reader should be aware that plumes arising from disposal and secondary re-suspension may be significant in some cases. Material (especially muddy material) which has been dredged and advected to a new location will have lower shear strength than the original material and is therefore more susceptible to re-suspension. The hydrodynamics of the new location may also be more (or less) severe, which will affect potential re-suspension and onward advection. The limits of the scope of this review are illustrated in Figure 1.2.

Box 1.1 *Definition of relevant terminology*

Acoustic profiler	An acoustic profiler transmits a beam of sound into the water column. Backscattered sound from particles in the water is received by the transducer. The technique can record continuous through-depth current velocities by using the doppler shift of the return signal to estimate the velocity of passing water. Moreover, the acoustic strength of the received signal can be used to measure suspended sediment concentrations.
Advection	The movement of sediment (the plume) with the current flow.
Aggregate(s)	Sand and gravel suitable for use in the construction industry for mixing with a matrix to form concrete, macadam, mortar or plaster, or used alone, as in railway ballast, unbound road-stone or graded fill.
Bathymetry	The level of the seabed.
Bed shear stress	The force per unit area applied to the bed by the action of water (waves and currents) flowing above it.
Clay	A fine-grained plastic sediment containing particles with a diameter less than 0.002 mm.
Coarse material	Cobbles, gravel and medium and coarse sand.
Coarse sand	Particles with diameter greater than 0.5 mm and less than 2 mm.
Cobbles	Particles with diameter greater than 60 mm and less than 200 mm.
Cohesive material	Sediments containing significant proportion of clay, the electro-chemical properties of which cause the sediment to bind together.
Critical bed shear stress for deposition	The point at which the force on the bed due to currents and waves is sufficient to prevent the settling of sediment on the bed.
Diffusion	The gradual spreading of the sediment plume as a result of turbulence.
Dispersion	The gradual spreading of the sediment plume through large-scale spatial current speed variations.
Dynamic plume	Jet of sediment-laden water that initially, after release from the dredger, moves rapidly downwards towards the bed, under the influence of gravity.
Fine material	Silt and fine sand.
Fine sand	Particles with diameter greater than 0.063 mm and less than 0.25 mm.
Fluid mud	A cohesive suspension with a concentration of 3–100 kg/m^3 that occurs as a result of rapid settling (or fluidisation by wave action). It is easily mobilised by sloping beds and tidal currents.
Gravel	Particles with diameter greater than 2 mm and less than 60 mm.
Medium sand	Particles with diameter greater than 0.25 mm and less than 0.5 mm.
Minipod	An autonomous bottom-lander that carries timed and passive sediment traps and an array of sensors capable of measuring near-bottom-suspended sediment.

Mud	Sediment made up of material within the silt and clay size bands. Can contain other particle sizes too, eg, sandy mud.
Overflow	The return of suspended sediment from the dredger hopper or barge to the sea via spillways (ship-side or centrally mounted).
Passive plume	The movement of suspended sediment under the action of currents.
Silt	Particles with diameter greater than 0.002 mm and less than 0.063 mm.
Re-suspension	The erosion of deposited sediment.
Settling velocity	The speed at which sediment particles settle under the combined forces of gravity, friction and buoyancy.
Turbulence	Small-scale temporal and spatial variation in current speeds and wave action.

1.2.2 Geographical scope

The study area for the project was effectively the operating area of the UK dredging industry: the countries bordering the North Sea. However, the geographic coverage of the report varies based on the issue under discussion. For example, the report is necessarily UK-focused in the context of legislation and stakeholders (although it takes account of interests in other countries), but draws on global experience regarding the dynamics of sediment plumes. This is important in the context of the industry in the UK, since it has strong links with Europe and other countries through multinational contractors, international dredging associations and legislation introduced by the European Community and international conventions.

1.2.3 Approach

The investigations undertaken for the scoping report included a review of the legislation affecting marine and coastal dredging relevant to sediment plumes and the identification of the current efficiency of knowledge transfer on sediment plumes between the industry's stakeholders. Investigations for the latter included a consultation exercise, involving various stakeholders from the UK and overseas dredging industry.
The core of the scoping assessment involved a review of research work on sediment plumes undertaken by the industry, government departments (ie regulators) and research organisations in the UK and overseas. This review covered the dynamics of sediment plumes in the water column and at the seabed, and the physical processes affecting sediment plumes. It also covered the environmental effects and potential impacts associated with sediment plumes. Mitigation methods to control the extent of sediment plumes were also considered.

Finally, the scoping exercise identified gaps in current knowledge on sediment plumes, suggested potential research initiatives to fill these gaps, and considered how to move forward in terms of obtaining good data and developing a comprehensive, multi-disciplinary assessment framework.

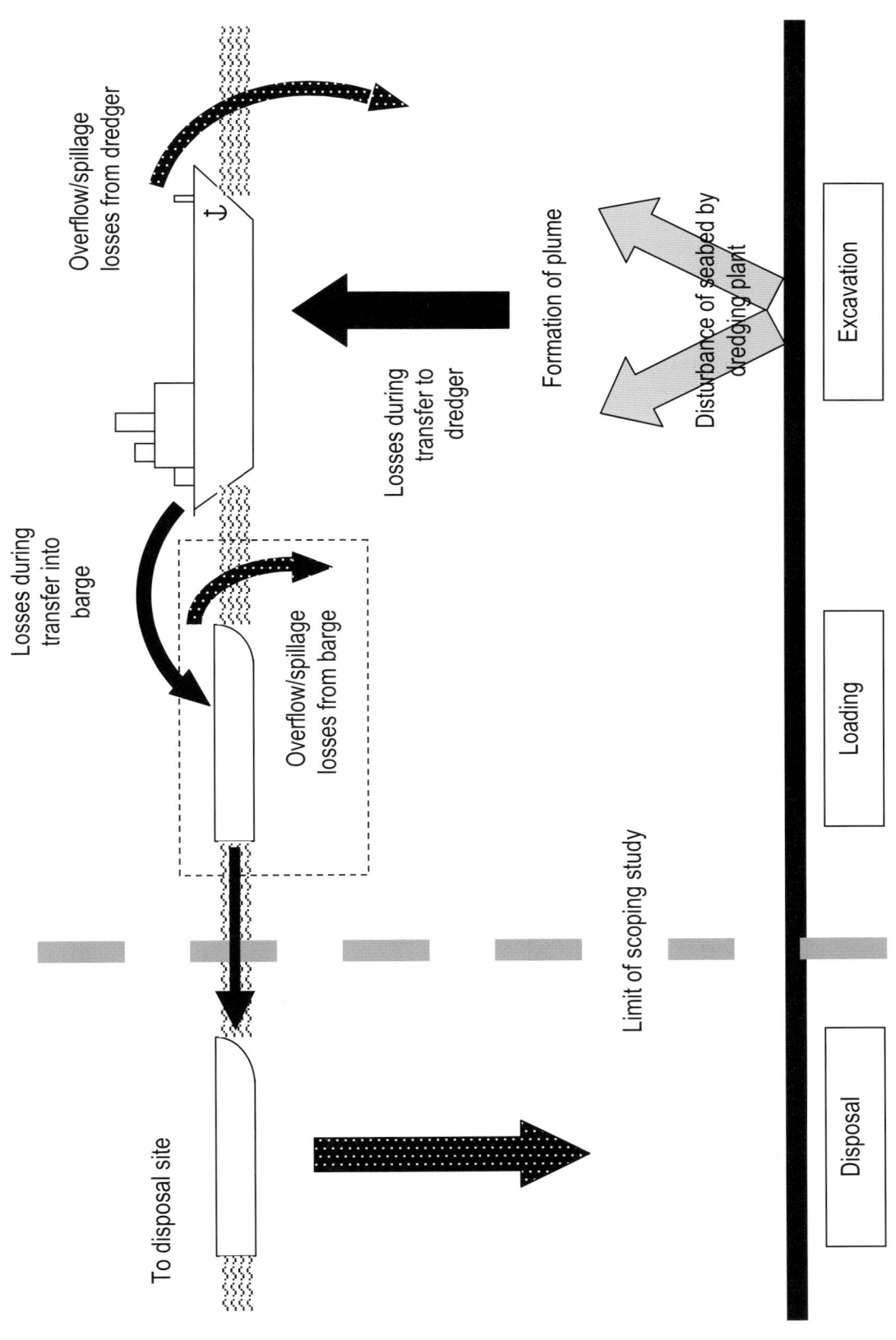

Figure 1.2 *Limits of the technical scope of the report*

1.2.4 A user's guide

The key components of this sequence, from the dredging project to its effect, are illustrated in Figure 1.1. As far as possible, this figure demonstrates the key issues – and thereby the route through the report – for marine aggregate and (separately) capital and maintenance dredging. However, the various facets of plume generation (ie the characteristics of each dredging project, the environment to be dredged, the dredger to be used and the plume that arises) are strongly interrelated in terms of the potential effect on the environment. Straight-line relationships are difficult to derive.

In meeting the objectives of the project, this scoping report sets the scene for the assessment of sediment plumes by outlining the legislative context within which plumes are regulated (Section 2.1 and Appendix 1) and by describing the stakeholders who participate in the process (Section 2.2 and Appendix 2).

Section 3 then describes the various sources of sediment plumes (the "source term') and the plume types that arise from dredging a variety of materials, in different environments, using several dredging techniques. It focuses, in particular, on the causes and magnitude of the sediment losses that result from the use of different dredging plant. Once in the water column, Section 4 discusses current knowledge of the processes affecting dynamic plumes and passive plumes (Sections 4.2 and 4.3 respectively) and the models available for predicting them (Section 4.4 and Appendices 5 and 6). Both Sections 3 and 4 draw on the results of recent research, field measure-ments (developed further in Appendix 3) and monitoring programmes (Section 3.3 and Appendix 4).

The scope of the potential environmental effects of sediment plumes is reviewed in Section 5, which focuses on water quality, marine ecology, fish and shellfish, and relates impacts, as far as possible, to the type of dredging project being undertaken. Section 6 then identifies the measures that can be taken to mitigate environmental effects.

Further details of the types of environmental effects typically associated with marine aggregate dredging and capital and maintenance dredging are provided in Box 1.2.

Sections 7 and 8 consider the way forward, with Section 7 outlining the requirements of an assessment framework (Section 7.1). From this and the preceding sections, the gaps in current knowledge are identified (Section 7.2). Section 8 offers recommendations on the further work required to produce guidance and develop a comprehensive framework for the assessment of sediment plumes arising from dredging.

Aggregate dredging

Capital and maintenance dredging

To distinguish between those issues and effects typically relevant to aggregate dredging in comparison to those relevant to capital and maintenance dredging, marginal flags have been used, as illustrated here. However, please note that while the associated text will generally be specifically applicable to the type of dredging in question, this will not be true in all cases. Aggregate, capital and maintenance dredging can all be undertaken using a variety of techniques, to dredge a variety of materials in a variety of environments. However, aggregate dredging in the UK is carried out almost exclusively by suction hopper dredgers, designed for the purpose, which operate in fairly consistent conditions and deposits. Nevertheless, there exist many possible scenarios, all of which must be considered in assessing the influence of sediment plumes.

Box 1.2 *Effects on the marine environment*

> The effects of aggregate and capital and maintenance dredging on the marine environment are considered in detail in Section 5 and are summarised below. Although there is the potential for similar environmental effects to arise due to different types of dredging, the likelihood of a particular effect occurring and the severity of the effect will often vary depending on the nature of the dredging project. The following indicates which effects typically occur in each case.
>
> | Aggregate dredging | Section | Capital and maintenance dredging |
> |---|---|---|
> | *Reduced water transparency* | 5.2.2 | *Reduced water transparency* |
> | *Contaminant mobilisation* | 5.2.3 | *Contaminant mobilisation* |
> | *Increased oxygen demand* | 5.2.4 | *Increased oxygen demand* |
> | Effect on seabed communities | 5.3.2 and 5.3.3 | Effect on estuarine and shallow littoral benthic communities |
> | Effects on juvenile and adult fish | 5.4 | Effects on juvenile and adult fish |
> | *Effects on fish migration* | 5.4.4 | *Effects on fish migration* |
> | Effects on shellfisheries | 5.5 | Effects on shellfisheries |
> | *Effects on sedimentology* | 5.6.1 | *Effects on sedimentology* |
> | Disruption of designated habitats/species | 5.6.2 | Disruption of designated habitats/species |
>
> **Note:** Italicised items indicate that, in general, only a limited risk exists

1.3 DREDGING IN THE UK AND WORLDWIDE

1.3.1 Introduction

Dredging is undertaken to deepen or extend harbour approach channels and berths (capital dredging), to maintain the bed of these channels and berths at their required level (maintenance dredging), and to remove sediments such as sand and gravel from the seabed to use as aggregate for the construction industry (aggregate dredging), and to gain material for beach recharge, land reclamation and infrastructural works.

The amount of dredging undertaken around the world varies enormously, with the magnitude of dredging operations in the UK being relatively large. According to the IADC (pers comm), global figures for dredging are not available owing to the lack of information regarding the productivity of different types of vessels. However, the UK figures presented in Box 1.3 can be put into context by considering in general terms the level of dredging activity ongoing in other countries. For example, the total annual dredging undertaken in the USA is approximately ten times higher than that of the UK, while the dredging undertaken in Portugal is approximately ten times smaller.

1.3.2 Aggregate dredging

Aggregate dredging has developed as a separate "industry" in the UK more than in most other countries. The plant used is almost exclusively the trailing suction hopper dredger (see Section 3.2.2) and most of the work is carried out offshore.

Aggregate dredging is predominantly undertaken to provide sand and gravel for use in the construction industry, but is also undertaken to provide materials for reclamation and beach recharge schemes. More infrequently, it can be used for habitat creation.

In order to be economical, the industry prospects for sites that will yield material of a suitable quality (with a minimum of processing) for sale either in the UK or other European countries. This means that operations are typically restricted to a particular type of environment, such as one in which the water depth is not too great and the bed material is sand or gravel with a minimum amount of fine material. In some cases the material is processed on board by washing out unwanted fine material and discharging it overboard (screening). Clearly this process can be a significant generator of plumes of suspended solids, depending on the fines content of the aggregates. Such dredging often takes place in high-energy conditions (ie environments characterised by strong currents and/or waves).

Box 1.3   *Dredging quantities in the UK*

> There are currently more than 100 dredgers operated by UK ports and dredging companies (Dredging and Port Construction, 1999). The best available figures for the quantities of maintenance and capital dredging arising in the UK relate to the period 1985 to 1994 (MAFF, 1995).
>
> Over this period the average annual total of material dredged for maintenance dredging in the UK was 24.4 million tonnes (gross hopper load) or 11.3 million tonnes (dry solids). The latest validated figures are for 1996 and come to an annual total of 38 million tonnes (gross hopper load, CEFAS, pers comm). The average total of material dredged for capital dredging in the UK over the same period was 4.7 million tonnes (gross hopper load) or 3.1 million tonnes (dry solids), although this figure has varied from year to year. The latest validated figures are for 1996 and come to an annual total of 13.5 million tonnes (gross hopper load, CEFAS, pers comm).
>
> Over the past decade the UK marine aggregate dredging industry for the extraction of sand and gravel has recovered an average of 24 million tonnes per annum (HR Wallingford, 1999a). Just over half of this resource is landed on the south and east coasts, representing one-third of the local demand for construction sand and gravel. The total contribution for England and Wales is about 12–13 per cent. Marine aggregates are becoming more important due to increasing constraints on land-based quarrying, and the UK Government is promoting greater use of marine resources.

1.3.3 Maintenance dredging

Maintenance dredging takes place in harbour berths, marinas and approach channels and usually involves the removal of silt and clay or sand.

Where maintenance dredging occurs in quiescent (low currents and waves) conditions, such as in sheltered harbour berths, the sediment will usually be mud. In higher-energy conditions, such as exposed approach channels, the sediment to be dredged is more likely to be predominantly sand.

Maintenance dredged silts and muds are typically deposited in designated disposal sites. However, they may also be used in beneficial use schemes to re-create intertidal habitat and, more recently, have been introduced back into the system (beyond the point at which they become "trapped") as part of sediment replacement initiatives in eroding systems. Beneficial uses are typically sought for maintenance-dredged sands.

A variety of dredging methods are used for maintenance dredging, including: trailing suction hopper, grab, backhoe, bucket ladder, cutter suction, hydrodynamic, and auger dredging. The choice of method used is generally based on economics but also depends on the nature of the dredging project, the physical constraints of the dredging location and the hydrodynamic environment. Each of these methods is introduced below and is described in detail in Section 3.2.

1.3.4 Capital dredging

Capital dredging also takes place mainly in harbour berths and entrances, and in the same way can occur in both quiescent (berths) and high-energy conditions (approach channels). While maintenance dredging is more concerned with removal of the top layer of sediment deposited since the last dredge, capital dredging removes previously undisturbed material. The nature of the material can vary from rock to mud.

By far the most common type of dredger used in capital dredging in the UK is the trailing suction hopper dredger. The reason for this is that the great majority of capital dredged material in the UK is placed at offshore disposal sites and the trailing suction hopper usually represents the most efficient dredging and transport solution in this situation. Where material is being pumped onto land, or to another location for disposal, the cutter suction dredger (details provided in Section 3.2.4) is a more common choice, especially in other European countries.

1.4 DREDGING TECHNIQUES

The following paragraphs briefly describe common dredging methods, their use for different dredging operations and the main mechanisms by which they release sediment and thus potentially create plumes. More detail regarding each of the dredging methods is given in Section 3.2. The list is not exhaustive and some types are in much more common use than others, as shown in the following table.

Table 1.1 provides the total world figures for each type of dredger in 1992. Current figures are expected to broadly reflect the same distribution. The figures show that cutter suction dredgers are by far the most numerous worldwide, with grab dredgers following cutters. However, many of the trailing suction hopper dredgers in operation are very large and tend to have a high utilisation rate, which means that in terms of the total quantity of material dredged they are more dominant than the table indicates.

Table 1.1 *World totals for dredger type (Bray et al, 1997)*

DREDGER TYPE	Trailing suction hopper	Cutter suction	Dustpan	Bucket ladder	Backhoe	Dipper	Grab	Others
Total number	318	1001	7	196	115	25	497	239

1.4.1 Trailing suction hopper dredgers

Trailing suction hopper dredgers are used for both aggregate dredging (almost exclusively) and capital and maintenance dredging, particularly where material is to be disposed of at an offshore site. This type of dredger pumps a water-sediment mixture (ie slurry) from the seabed to an onboard hopper via a suction pipeline. Once in the hopper, the coarse sediment settles to the bottom and the supernatant water is returned to the sea via an overflow weir. The term "overflow" describes this process, which can be a significant source of sediment plume generation. Furthermore, since the residence time in the hopper is short (decreasing as the hopper fills), much of the fine fraction of the sediment does not settle out and is released into the water with the overflow discharge.

1.4.2 Suction dredgers

Suction dredgers are largely used for aggregate dredging. They share some of the characteristics of trailing suction hopper dredgers, but are stationary while dredging. The potential for sediment losses is similar to those for trailing suction hopper dredging.

The main feature of a suction dredger is that it sticks its suction pipe deep into the ground, remaining stationary. Sand has to flow from a relatively high face to the suction mouth, leaving deep holes or troughs in the seafloor. In general, suction dredgers load barges or pump straight ashore. This type of suction dredger is used in international dredging. Some suction dredgers can load sand into their own hopper. This type is used for aggregate dredging in the UK, and internationally for dredging small rivers.

1.4.3 Cutter suction dredgers

Cutter suction dredgers are used for both capital and maintenance dredging projects, particularly where stiffer cohesive sediments and weak rocks need to be dredged, or when it is appropriate to pump the material ashore (eg for reclamation). The dredging process entails the use of a rotating cutterhead mounted on the end of a suction pipeline. The material is dislodged by the cutter head and pumped, as slurry, via a pipeline to the desired location. Unlike some other techniques, there is no overflow from the dredger. The main source of sediment re-suspension is material disturbed around the cutterhead. It is not usual practice to pump the slurry directly into barges for onward transport, but in such cases the potential for losses due to splashing and overflow are large.

1.4.4 Dustpan dredgers

Dustpan dredgers are usually suction dredgers (without their own hopper) which use a wide suction mouth to collect a wider but thinner deposit layer. They are often used to remove fresh deposits from areas proposed for capital works prior to the placement of pipelines and tunnels, and for maintenance dredging on rivers.

1.4.5 Grab dredgers

Grab dredgers are used for dredging cohesive and non-cohesive sediment during capital and maintenance dredging operations, especially within confined areas such as harbours. The dredgers typically comprise a clamshell grab fixed to a crane. Sediment re-suspension occurs when the grab impacts with the seabed, during bed disturbance when material is initially removed, and by spillage as the grab is hoisted or lowered through the water column and its contents are loaded into a barge.

1.4.6 Backhoe dredgers

Backhoe dredgers are used for dredging cohesive and non-cohesive sediment during capital and maintenance dredging and are also used for dredging in confined seabed areas, although they are no longer being made. Backhoe dredgers are similar to land-based excavators, although during dredging the excavator is usually mounted on the end of a pontoon, which has to be fixed into position using spikes (known as spuds) pushed into the seabed. Like grab dredgers, backhoe dredgers re-suspend sediment when the bucket hits the seabed, and because of spillage as the bucket is lifted or lowered through the water column and when its contents are loaded into a barge.

1.4.7 Bucket ladder dredgers

Bucket ladder dredgers are used for dredging cohesive and non-cohesive sediment during capital and maintenance dredging operations. Sediment is removed by a continuous chain of buckets that scoop material from the bed. The sediment is raised through the water column and deposited onto chutes as the buckets pass over the top of the ladder. The chutes convey the sediment to a hopper. Sediment is released into the water column due to bed disturbance by the buckets, spillage from the buckets, leakage from the chutes and during the loading of barges.

1.4.8 Hydrodynamic dredgers

Hydrodynamic dredging is used, relatively infrequently, for maintenance dredging (typically in estuaries). There are two main classes: those that deliberately inject the sediment into the flow at high energy to obtain maximum dispersion and advection, and those that inject water into the bed sediment in order to create a flow of sediment/water mixture at the bed away from the dredging area. The former technique can be adopted using most types of dredging plant and, by definition, creates high concentration sediment plumes. The latter is usually termed "water injection dredging" and, depending on the operating conditions and the nature of the sediment, may give rise to relatively low concentrations of suspended sediment above the sediment flow.

1.4.9 Environmental dredgers

Environmental dredgers are designed to reduce sediment losses and/or to enable greater accuracy when dredging contaminated sediments. They are specialised dredging plant and so are not commonplace and used only relatively infrequently. Environmental dredging is adopted during capital and maintenance operations involving contaminated sediments to minimise re-suspension, and thereby control the redistribution of contaminants. Such dredgers are based on conventional dredging systems that have been modified for minimising sediment re-suspension. Most types rely on a special covered cutter or drag head to control sediment release into the aquatic environment.

1.4.10 Barge loading

When discharging into barges, any of the above plant may be a source of sediment plumes due to splashing and overflow. Dredgers that have built-in hoppers can create plumes in the same way.

1.5 SEDIMENT PLUMES

1.5.1 Introduction

This section provides a short introduction to the nature of plumes caused by different dredging practices. Further details are provided in Section 3 (Sources of sediment plumes) and Section 4 (Plume processes and modelling).

Dredging releases sediment into the water column, forming a sediment plume, of which there are two types, or phases; the dynamic phase (where the plume moves under its own volition) and the passive phase (where, broadly, the plume moves due to other influences acting upon it). In the dynamic phase, plume behaviour is mainly determined by the nature and concentration of the material and how it is placed into the water. The main causes of such plumes are trailing suction hopper dredger overflow, screening during aggregate dredging, pipeline discharge in the aquatic environment and hopper discharge either through bottom opening or pumped discharge.

In the passive phase, plume movement is controlled to a greater extent by the hydrodynamic environment (mainly the strength and direction of the current). Sources include losses into suspension during the dynamic plume phase and various actions that cause the plume to respond or "move" during dredging operations.

Though it is useful, both for understanding and predictive modelling, to separate dynamic and passive plumes into two distinct phases, in practice there is no precise distinction between them. Furthermore, there is a (small) dynamic aspect to all passive plumes due to the increased density of the plume above that of the surrounding waters.

1.5.2 The dynamic plume phase

The introduction of material into the water column results in a water/sediment mixture of higher density than the surrounding water, and it therefore descends towards to the seabed. In the case of hopper or barge overflow, the mixture is likely to have additional downward momentum because of the mode of release. This initial rapid descent of the plume is referred to as the "dynamic phase" of plume dispersion (Figure 1.3). A similar process exists with disposal but disposal is outside the scope of this report.

As the plume descends, a proportion of the sediment (usually a small proportion) is stripped from the plume into the surrounding water column and advected away from the dredging area by currents as a passive plume (see Section 1.5.3). The remainder of the released sediment impacts upon the bed. Some sediment is re-suspended as a result of the impact, while the remainder moves radially outwards across the seabed as a dense pancake-like plume, slowing with time and distance as the kinetic energy is spent overcoming friction. The flow can be modified by both bed slope and currents. Eventually a weak deposit is formed (HR Wallingford, 1999b). Re-suspension can also take place during this radial flow phase.

The zone of impact of the dynamic plume is relatively small, usually affecting an area less than 100–200 m from the dredger. The size of the impact zone is principally dependent on the initial density and momentum of the sediment/water mixture and the strength of the current flow. The suspended sediment concentration within a dynamic plume is higher than that within a passive plume; it can be several thousands of milligrams of sediment per litre of water (mg/l).

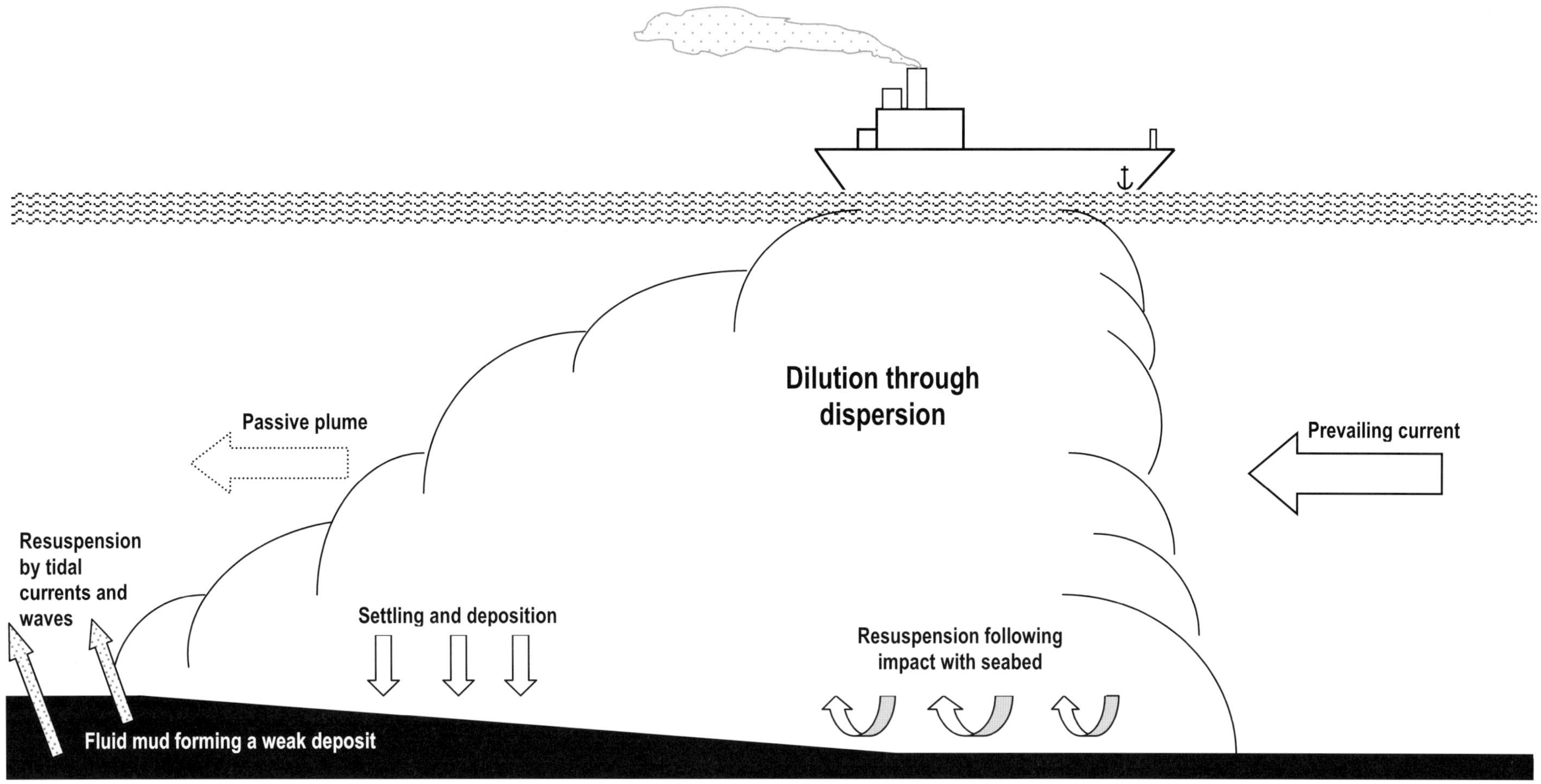

Figure 1.3 *Dynamic plume phase (due to overflowing)*

1.5.3 The passive plume phase

Passive plumes may be generated by the dynamic plume phase and by different types of dredging operation. A passive plume generated by a dredging operation is shown in Figure 1.4. The sediment concentrations within a passive plume are relatively low and the settling velocity of the particles is sufficiently low for them to remain in suspension for a significant time. For the finest particles this may mean several hours or even days. The movement of the particles is thus dominated by the water currents. If the current speeds are high, there may be sufficient turbulence to keep the sediment in suspension indefinitely, although in the case of tidal currents there are periods when the currents are low enough to allow settlement. The zone of settlement to the bed is often called the "footprint", the horizontal movement is "advection" and the process whereby the plume spreads in width and depth is termed "dispersion" or "diffusion", depending on the scale of the mixing process.

In the case of generation due to the dynamic phase, a passive plume forms when sediment is stripped from the plume into the water column during the rapid descent of sediment, as a result of the impact of the dynamic plume on the bed, or subsequently during the flow of sediment along the bed.

There are two main mechanisms that cause this to occur:

- turbulence (the small-scale temporal and spatial variations in current flow). This mechanism is referred to as diffusion
- the effect of different current velocities through the water column, which results in particles at different heights travelling in different directions and at different speeds, thus spreading the plume. This mechanism is referred to a dispersion.

Other sources of a passive plume are the losses arising during dredging operations. For example, in the case of a grab dredger, sediment may be re-suspended as the grab hits the seabed, as it closes, as it is raised through the water, as it is slewed and as it is lowered through the water. Each operation is discussed in this respect in Chapter 3.

The zone of influence of the passive plume can be several kilometres or more, and is dependent on the magnitude and direction of tidal currents and the magnitude and nature of the sediment released. Suspended sediment concentrations within the plume can be in the order of hundreds of mg/l in the vicinity of the dredger, reducing to tens of mg/l with distance from the dredger.

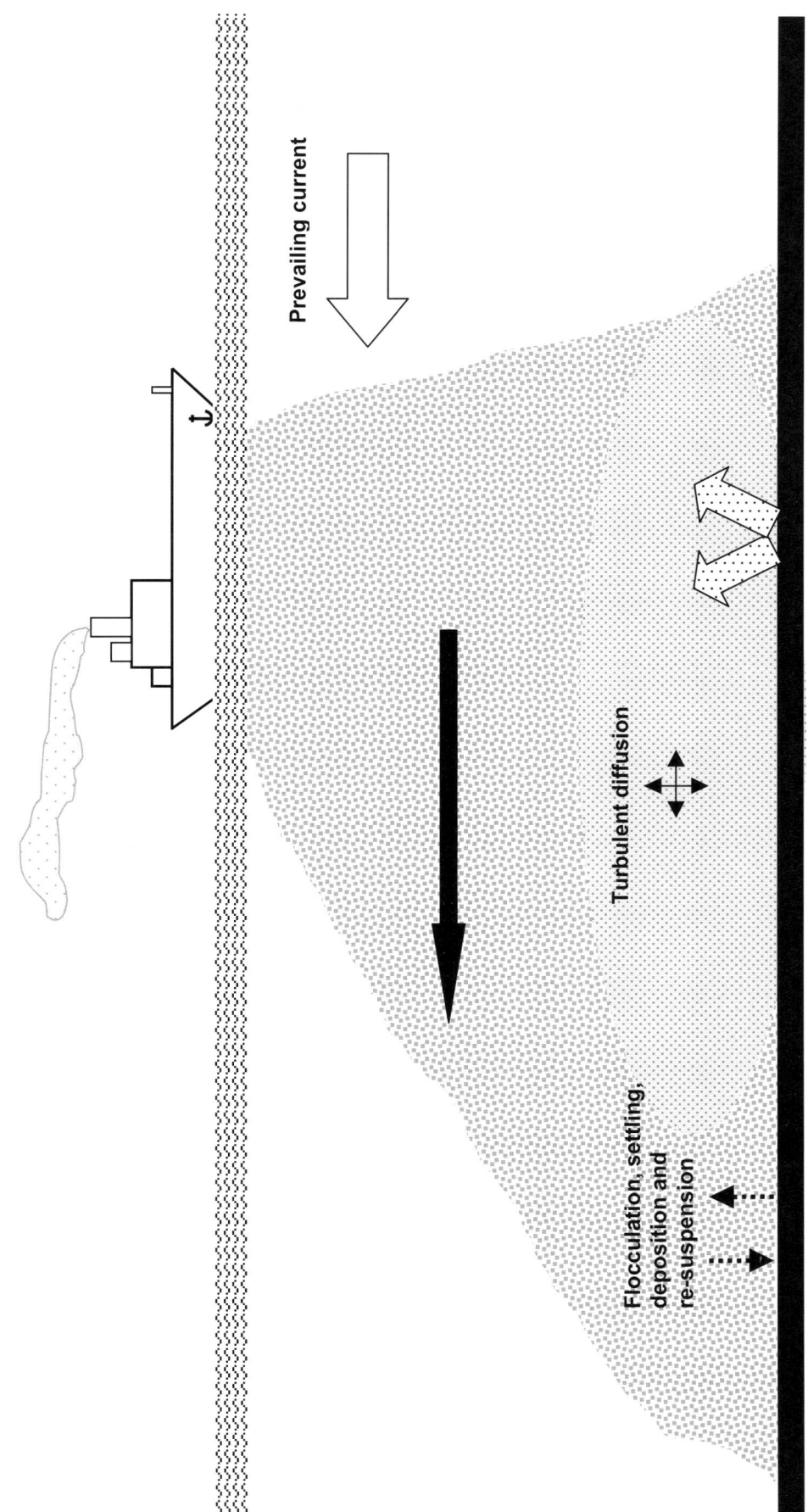

Figure 1.4 Passive plume phase

CIRIA C547

2 Legislation, regulation and stakeholders

This chapter summaries the legislative and regulatory systems affecting marine aggregate dredging and capital and maintenance dredging in the UK and, where relevant, overseas. The UK does not have any legislation specifically relating to sediment plumes. As a result, sediment plumes are usually considered in the context of environmental legislation relating to the licensing and permitting of all aspects of dredging projects. In most cases, the requirement to investigate sediment plume effects is implicit in the EIA process (where applicable).

This chapter also introduces the stakeholders with an interest in sediment plumes, including the industry, regulators and other affected (or potentially affected) parties. The primary interactions between stakeholders are shown, generically, on figures accompanying the text.

Since sediment plumes are not explicitly or independently linked to legislation, regulation and stakeholder interests, this section is deliberately brief. More detail about legislation and regulation is, however, presented in Appendix 1 and further details on stakeholders are provided in Appendix 2.

2.1 LEGISLATION AND REGULATION

2.1.1 Marine aggregate dredging

Legislation and regulation in the UK

Marine dredging of aggregate for commercial purposes in the waters of England and Wales will be statutorily controlled by the government under the provisions of the forthcoming Environmental Impact Assessment and Habitats (Extraction of Minerals by Marine Dredging) Regulations 2000. Such dredging in Scottish waters is statutorily controlled by Scottish ministers under the Environmental Impact Assessment and Habitats (Extraction of Minerals by Marine Dredging) (Scotland) Regulations 1999, which will come into force in 2000.

These statutory regulations will replace the informal government view procedure that has operated since 1968. The procedure prevented the Crown Estate from granting any dredging licences unless the government was satisfied that the dredging would not harm the environment. The Crown Estate would not licence any such dredging until it received a favourable government view.

The regulations will transpose into UK legislation, in so far as marine dredging is concerned, the provisions of EC Directive 85/337/EEC, as amended by EC Directive 97/11/EC, on the assessment of the effects of certain public and private projects on the environment as well as provisions of EC Directive 92/43/EEC on the conservation of natural habitats and of wild fauna and flora (the Habitats Directive).

Once the regulations come into force, an operator proposing to carry out dredging can request the regulator (see Section 2.2.2) to determine whether the proposal is likely to have a significant effect on the environment by virtue of its nature, size or location. If the regulator determines that there will be a significant impact, the operator will be required to apply to the regulator for a dredging permission. An environmental statement and a coastal impact study, as well as a statutory dredging application fee, must accompany such applications.

The regulator will determine dredging applications after extensive consultation and will either refuse or permit them subject to conditions to protect the environment. Owners of the seabed will not usually issue a commercial licence to a dredging company unless dredging permission is first obtained. It will be a criminal offence for any dredging to be undertaken either not in accordance with an agreement between the operator and the owner of the seabed entered into before the regulations come into force; or without the benefit of dredging permission granted by the regulator or a statement by the regulator that dredging permission is not required.

Decisions by the regulator on dredging proposals and the determination of dredging applications will, therefore, be completely unrelated to the need for an operator to obtain either a licence or some other form of agreement from the owner of the seabed for dredging to be undertaken.

The consideration of sediment plumes in the determination of dredging applications will not be specifically required under the new regulations. However, as the preparation of the environmental statement to accompany all such applications will require all environmental aspects to be considered, in the context of aggregate dredging, the need to address environmental implications of sediment plumes is implicit.

The consideration of such environmental effects will be particularly important if the proposed dredging is either within, adjacent to, or otherwise likely to affect a European protected site of nature conservation importance (eg a Special Area of Conservation or SAC).

Legislation and regulation overseas

Western Europe has legislation and regulatory processes in place regarding the licensing of aggregate dredging that are similar to that implemented in the UK. Legislation for Denmark, the Netherlands, Germany and France is briefly described in Appendix 1.

In the United States, offshore sand and gravel extraction is regulated by the Department of the Interior's Minerals Management Service (MMS) under federal law. A regulatory system is in place to support all policies relating to prospecting, environmental analysis, leasing and operations. Environmental consideration is enforced further through the National Environmental Policy Act.

2.1.2 Capital and maintenance dredging

Legislation and regulation in the UK

Like aggregate dredging, sediment plumes arising during capital and maintenance dredging are only addressed by the regulatory process if they are considered to be likely to have an environmental effect as part of an EIA.

The legislation primarily relating to the environmental impact of capital and maintenance dredging is:

- Harbour Works (Environmental Impact Assessment) (England and Wales) Regulations 1999
- Town and Country Planning (Environmental Impact Assessment) (England and Wales) Regulations 1999
- Environmental Impact Assessment (Scotland) Regulations 1999.

The UK implements provisions for EIA made under EC Directive 97/11/EC through various regulations, including those listed above (and their future amendments). This legislation introduces both mandatory and discretionary requirements for EIA (based on thresholds) through statutory permissions and consents, regulated by the DETR, when harbour works are proposed and require harbour empowerment/revision orders and/or Coast Protection Act consent. EIAs for port and inland waterway development are also required in order to apply for planning permission. There is separate legislation for England and Wales and for Scotland. Appendix 1 provides more details.

Again, the EIA process does not specify the examination of sediment plumes, but it does require consideration of the environmental effects of all aspects of a proposed project, including the sediment plume.

Similarly, if capital and maintenance dredging is proposed within or adjacent to a European protected site then the provisions of EC Directive 92/43/EEC (the Habitats Directive) are applied in the UK through the Conservation (Natural Habitats etc) Regulations, 1994. In this case, as for aggregate dredging, if it is considered that the works are likely to have a significant effect on the European site, then the competent authority for licensing the dredging is required to make an "appropriate assessment" of the affect of the dredging (including the plume) on the nature conservation status of the site; a decision which the applicant is required to inform. As before, if the assessment determines that the works are likely to adversely affect the integrity of the site, they will not be able to proceed without mitigation or, following agreement of overriding public interest, compensation.

The Food and Environment Protection Act 1985 (FEPA), which gives MAFF control over the deposit of dredged material at sea, does not control any part of the dredging process.

Legislation and regulation overseas

European countries that are member states of the European Union are subject to the provisions of EC directives (as detailed above); although each country may implement these provisions in different ways.

In Hong Kong, by comparison, the Environmental Impact Ordinance is established to avoid, minimise and control the adverse environmental impacts of designated projects through the EIA process and environmental permits. Permits are required for projects involving reclamation works (including associated dredging) and dredging operations exceeding 50 000 m^3. Both types of project can involve capital and maintenance dredging. To date, the Environmental Protection Department has issued several dredging permits. All permits include detailed monitoring and auditing conditions on dredging operations to prevent adverse environmental effects associated with the introduction of sediment into the water column. Further details are provided in Appendix 1.

2.2 STAKEHOLDERS

In the context of sediment plumes, stakeholders include all of those parties who are affected by their generation, that is, the industry that causes them to arise (and their representative bodies), the agencies that are responsible for their regulation, the companies, institutions and individuals that investigate them, and the individuals and organisations whose interests are (or are perceived to be) affected by them (eg fishermen). To a significant extent, the stakeholders' interest in sediment plumes is driven by the requirements of the legislation described in Section 2.1 and Appendix 1. Accordingly, and particularly with respect to the regulators, their direct association with sediment plumes is tenuous in so much that they are associated through the need to consider the environmental effects of dredging in general, and therefore sediment plumes, in their commercial or regulatory responsibilities.

The following sections identify the dredging industry's stakeholders. Figure 2.1 shows the primary interactions between the UK stakeholders involved in aggregate dredging and capital and maintenance dredging. Further details on the key organisations described below and the transfer of knowledge between them are included in Appendix 2.

2.2.1 The dredging industry

Many international dredging companies and contractors carry out both aggregate dredging and capital and maintenance dredging. However, in the UK, the extraction of marine aggregates tends to be undertaken by UK companies that solely supply the land-based construction industry and beach nourishment projects. Capital and maintenance dredging for ports and other maritime projects, on the other hand, tends to be carried out by multinational dredging contractors. The situation in Europe is different, where aggregate dredging and capital and maintenance dredging are not separate industries. Many of the dredging industry's interests are looked after by various associations, such as those detailed below.

British Marine Aggregate Producers Association (BMAPA)
BMAPA was established to increase awareness of the marine aggregate dredging industry and to raise the profile and promote the interests of the aggregate dredging industry. BMAPA supports an aggregate dredging R&D programme.

Federation of Dredging Contractors (FDC)
The FDC is a trade association for major dredging contractors. Membership is open to those contractors with an operating office in the UK and the capability to work internationally. The Federation's objective is to represent the interests of members on matters of common concern.

International Association of Dredging Contractors (IADC)
The IADC, based in the Netherlands, is a worldwide umbrella organisation for more than 120 private dredging companies. It seeks to promote the private dredging industry and establish fair, open market conditions for its members.

Central Dredging Association (CEDA)
CEDA is based in the Netherlands and has British, Belgian and Dutch national sections that promote education about dredging, disseminate quality information on dredging, develop guidelines and standards relating to good practice, initiate and support research, and be proactive in relevant policy-making.

Vereniging van Waterbouwers in Bagger-, Kust- en Oeverwerken (VBKO)
The VBKO is the Dutch association of contractors that undertake dredging, shore and bank protection works, representing more than 300 companies, and provides an outlet for education and business promotion.

Permanent International Association of Navigational Congresses (PIANC)
PIANC, based in Belgium, represents organisations involved in the safe and efficient operation of all types of commercial and recreational vessels, including the management and sustainable development of navigational waterways.

2.2.2 Regulators

The new regulations described in Section 2.1.1 define as regulators the Secretary of State for the Environment, Transport and the Regions, Scottish ministers, the National Assembly for Wales and the Department of the Environment (Northern Ireland) (DOENI). They have responsibility for the statutory control of marine aggregate dredging within the waters of, respectively, England, Scotland, Wales and Northern Ireland.

Under the regulations described in Section 2.1.2, capital dredging is subject to a statutory order or consent under the Harbours Act 1964 or the Coast Protection Act 1949. Order and consent requirements depend on the circumstances of a specific dredging project. In the future, it is likely that the Secretary of State for the Environment, Transport and the Regions, Scottish Ministers, the National Assembly for Wales and the DOENI will be given responsibility for the statutory control of orders and consents within their waters.

In addition, other government agencies have regulatory responsibilities for dredging projects, as described in Appendix 2. These agencies include:

- Ministry of Agriculture, Fisheries and Food (MAFF) – responsible for advising the DETR on issues such as marine biology and fisheries
- Centre for Environment, Fisheries and Aquaculture Science (CEFAS) – providing advice to MAFF on the environmental effects of aggregate dredging
- Sea Fisheries Inspectorate (SFI) – advising MAFF on issues regarding commercial fisheries interests
- English Nature, the Countryside Council for Wales (CCW), Scottish Natural Heritage (SNH) and the DOENI – responsible for advising the Secretary of State for the Environment, Transport and the Regions about the potential environmental consequences of capital and maintenance dredging through EIA procedures, and aggregate dredging where statutory requirements might be compromised. In particular, the conservation agencies advise on the potential effects of aggregate dredging on marine SACs and the effects of capital and maintenance dredging on all European protected sites. They also provide guidance on meeting the objectives of Biodiversity Action Plans, which include habitats and species potentially affected by sediment plumes (eg honeycomb reef worms *Sabellaria* spp.).

2.2.3 Other interested parties

Other interested parties include port and harbour authorities, seabed owners (eg the Crown Estate), fishermen's associations, non-governmental organisations, members of the public, environmental consultants and coastal process specialists.

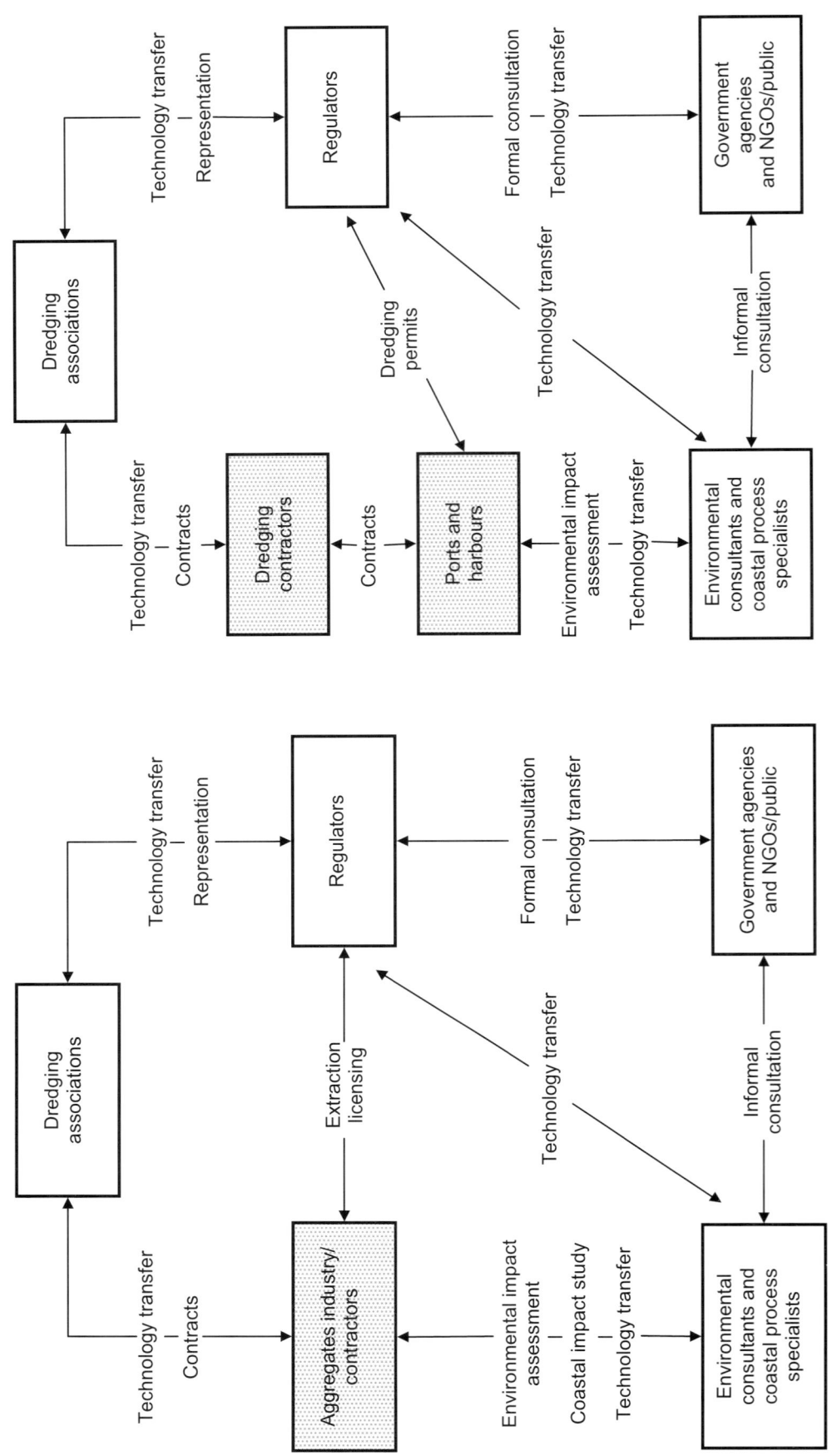

Figure 2.1 *Primary interactions between stakeholders in the UK aggregate dredging industry (left) and capital and maintenance dredging industry (right)*

2.3 KNOWLEDGE TRANSFER BETWEEN STAKEHOLDERS

There are many media by which knowledge on dredging issues, including sediment plumes, is transferred. For example, knowledge is shared through seminars and conferences, trade journals and published guidance, research and best practice. Perhaps the Internet represents the best recent approach to technology transfer. The role of different stakeholders in this process is introduced below and further details are provided in Appendix 2.

Knowledge transfer in the UK

There is scope for more direct knowledge transfer between dredging companies, although this might affect commercial competitiveness. Instead, the dredging associations listed in Section 2.2.1 serve as a focal point for information gathering and knowledge transfer. Typical approaches to transferring knowledge include the publication of books and magazines, attendance at and organisation of conferences, industry/association working groups, and the sponsorship of dredging seminars. It is these types of activities that bring together experts in the subject of dredging and present focal points for the transfer of knowledge on issues concerning dredging, including issues associated with sediment plumes.

Research is also carried out by the regulators, other government agencies and research institutions. For example, CEFAS monitors aggregate dredging sites in detail; the DETR and the Crown Estate both have R&D programmes which are concerned with dredging activity and, specifically, marine sand and gravel extraction; and the conservation agencies have commissioned research into the individual and combined environmental effects of aggregate dredging, including the effects of sediment plumes, on European sites of nature conservation importance (PDE, 2000). In addition, MAFF commissions research through its R&D initiatives into issues of relevance to sediment plumes. The Marine Aggregate Extraction Programme (AE09), for example, is concerned with the impact of mineral extraction on fishing activity and biota, and includes the following research projects of relevance:

- impact of dredger plumes on Race Bank and surrounding areas
- extraction of marine aggregates – investigation of marine environmental impacts
- beneficial use of dredged material
- role of the gravel biotope in the marine ecosystem and implications for the management of aggregate extraction, and
- a computer modelling tool for predicting the dispersion of sediment plumes from aggregate extraction activities.

Similarly, research initiatives of relevance are progressed under FEPA, which include consideration of the plumes arising due to sediment disposal, and by the Flood and Coastal Defence Emergencies Division of MAFF. This research is extensive and disparate. A CD-ROM is, therefore, currently being produced that will encompass the last ten years of MAFF-funded research.

Although such research is undertaken, it is typically published on an ad hoc basis, which limits access. However, there is an increasing use of the Internet to disseminate information, including the Marine Sand and Gravel Information Service (MAGIS) (see below) and UK studies for the US Minerals Management Service (see below).

Knowledge transfer overseas

Many of the dredging associations representing the industry in the UK are European-based international organisations that have connections further afield. Knowledge transfer on sediment plumes, therefore, has the potential to extend globally.

Recently, Dutch dredging companies and the Rijkswaterstaat, through the VBKO, have commissioned UK investigations into turbidity assessment software, the results of which will be in the public domain.

European regulators and the dredging industry also co-operate in various information programmes such as MAGIS, a data support initiative that involves dredging contractors, government licensing agencies and dredger manufacturers.

The Internet is perhaps the most efficient method of sharing information about sediment plumes. In Europe, the MAGIS programme, involving the DETR and the Crown Estate, includes an Internet site to disseminate information on aggregate dredging. The United States extensively uses the Internet to disseminate government information, an approach accelerated by statutory duties under the Freedom of Information Act. In terms of dredging and the sediment plumes associated with dredging, two US government agencies in particular provide vast amounts of publicly accessible information through the Internet:

- the Waterways Experiment Station of the US Army Corps of (http://chl.wes.army.mil/research/dredging/)
- the Minerals Management Service of the US Department of the Interior (http://www.mms.gov/intermar/environmentalstudiespage.htm).

Specific details on the dredging-related activities of USACE are provided in Box 2.1.

Box 2.1 *US Army Corps of Engineers*

> As a US government agency, the USACE has to make its research and technology publicly available. Most of its knowledge regarding dredging is made available through the Waterways Experiment Station (WES). The type of knowledge obtainable through the WES includes numerical models, regulatory guidance and research papers.
>
> In addition to providing access to free copies of published reports, all WES information is directly available through the Internet or via email. Information is readily accessible and easily transferable through the Internet. For example, the home page of the WES's Environment Laboratory includes a section dedicated to *Technology and information transfer*, providing links to sites for computer software, databases and publications. By accessing the publications link, all WES publications are available through further Internet links. A direct link to the DOER's technical notes provide access to publications such as *DOER-E5 Evaluation of dredged material plumes physical monitoring techniques* (Reine et al, 1998) and *DOER-E6 Estimating dredging sediment re-suspension sources* (DOER, 1999). Alternatively a direct link to the DOTS technical notes on the environmental effects of dredging provides access to publications on dredging plant such as *EEDP-09-1 Guide to selecting a dredge for minimising re-suspension of sediment* (DOTS, 1986) and *EEDP-09-2 Sediment re-suspension by selected dredges* (DOTS, 1988).
>
> The USACE also operates the Long-Term Effects of Dredging Operations (LEDO). LEDO has its own website via the WES and promotes technology transfer by leading the researcher into options for technical reports, technical notes, journal articles and conferences and presentations.

3   Sources of sediment plumes

3.1   INTRODUCTION

The sources of sediment plumes are essentially the losses, deliberate and otherwise, that occur during a dredging operation. There are three primary influences on the generation of sediment plumes: the dredging operation, the material and the hydrodynamic conditions in which the dredging takes place.

3.1.2   The dredging operation

Dredging operations are classified here as aggregate, capital and maintenance dredging, as discussed in Section 3.2. The dredging operation and its type defines the geographical location where the plume forms. The mechanical disturbance applied to the sediment and the locations associated with the dredging plant determine where losses (accidental or deliberate) can occur (eg spillage) and their magnitude. Each dredging technique is considered in turn in Section 3.2.

3.1.2   The material

The properties of the material to be dredged influence the amount and size distribution of the sediment that is released into suspension. The source strength, measured in terms of concentration by weight, of a dredged plume is highly dependent on the type of sediment being dredged, in particular its particle size distribution and the degree to which it disaggregates when disturbed by dredging. The actual amount of disaggregation is also clearly dependent on the amount of energy put into the dredging operation.

The material type also affects the turbidity of the plume because of the different optical properties of silty and sandy water. The finer the sediment, the higher the turbidity for any given concentration by weight. The influence of material type on the characteristics of the plume is discussed in Chapter 4. In brief, it should be noted that because of the different settling properties of sand and silt, the larger the proportion of fine sediment in the dredged material, the longer the resulting plume will take to settle out. This means that, in general, siltier materials produce more noticeable, longer-lasting plumes, although this typically does not equate to a greater impact.

The nature of the material to be dredged can vary considerably within a single dredging area. This includes areas licensed for aggregate dredging. Measurements taken during aggregate trailing suction hopper dredging (MMS, 1998), for example, show that overflow releases of fine sediment during aggregate dredging can vary within a single site by an order of magnitude.

3.1.3   The hydrodynamic conditions

The hydrodynamic environment affects how the dredging operation is carried out and therefore the rate of sediment loss. Maintenance and capital dredging can occur in both sheltered (for instance harbour berths) and high-energy conditions (eg entrance channels), aggregate dredging usually occurs in the open sea or large estuaries where high-energy conditions prevail.

Hydrodynamic conditions affect the choice of dredging plant. For example, cutter suction dredgers cannot operate in high wave conditions, whereas trailing suction hopper dredgers are versatile in this respect. It may be surmised that losses will generally be higher if operating conditions are difficult. This aspect should be taken into account in any attempts to model sources of plumes.

Quantification of losses

In considering the quantification of losses (sources), it is important to be aware of the way these have been observed and measured up to now. Section 3.3 discusses this in some detail, because of its importance in assessing sediment plumes, and gives examples of the results obtained. Reference is made where appropriate to field studies and a summary is provided in Section 3.4. More detailed accounts of the field studies are given in Appendix 3.

The factors affecting the plume and the plume dynamics after it has formed are discussed in Chapter 4.

3.2 EFFECT OF DREDGING TECHNIQUE ON SOURCE STRENGTH

This section describes the possible sources of sediment release from dredgers, beginning with a general review of the main differences between each class of dredging.

3.2.1 Influence of the type of dredging project

Aggregate, capital and maintenance dredging entail the use of various dredging methods, in different environments and the dredging of several types of sediment, and so exhibit differences in the type of plumes that they produce. The most common combinations are shown in Table 3.1 below.

Table 3.1 *Summary of the characteristics of different dredging projects*

Characteristics of the dredging operation	Aggregate dredging (UK)	Maintenance dredging	Capital dredging
Dredger	Trailing suction hopper dredger	All types	All types
Environment	Open sea	Harbour berths	Harbour berths
	Large estuaries	Entrance channels	Entrance channels
Material type	Sand and gravel	Mud, sand, gravel	All types
Source of sediment plume	Overflow	Overflow (TSHD)	Overflow (TSHD)
	Screening	Disturbance of bed (all types)	Disturbance of bed (all types)
	Disturbance of bed		
		Pipe leakage (CSD and all others using pipe transport)	Pipe leakage (CSD and all others using pipe transport)
		Agitation of bed (hydrodynamic)	Agitation of bed (hydrodynamic)

TSHD trailing suction hopper dredgers
CSD   cutter suction dredgers

Aggregate dredging

Aggregate dredging for sands and gravels mainly uses trailing suction hopper dredgers for which overflow, is acknowledged to be the dominant source of release of fine sediment (Bray *et al*, 1997; see also the summary of field data in Appendix 3). A description of losses specific to aggregate dredging is given in Section 3.2.2, describing losses during trailing suction hopper dredging.

Capital dredging

Capital dredging can involve a variety of dredging methods, the choice of which depends upon the dredging project, the material to be dredged (which can vary from soft mud to hard rock), the options for disposal and the hydrodynamic conditions at the site in question. Where capital dredged material is to be deposited offshore, as is usually the case in the UK, dredging is often undertaken using trailing suction hopper dredging because of the ability of this type of dredger to both dredge the material and transport it to the disposal site. Where disposal is to land, or where the dredging is of hard rock or clay, pipeline transport from cutter suction dredgers is more common.

Maintenance dredging

For maintenance dredging the types of material encountered are normally silts, sand and, to a lesser extent, gravel. In general terms only, it can be said that a typical maintenance dredging operation will put more fine sediment into suspension than a capital dredging operation of equal volume. Otherwise, the comments above regarding dredging method apply.

3.2.2 Trailing suction hopper dredging

Aggregate, capital and maintenance dredging

Trailing suction hopper dredgers (as illustrated in Figure 3.1) are used to dredge all but the strongest materials, but are not suitable for use in restricted areas. Hopper sizes vary from 750 m^3 to the 33 000 m^3 hopper of the *Vasco Da Gama* under construction in the Netherlands. Hopper sizes of under 3000 m^3 are the most common (Bray *et al*, 1997). As they move forward, trailing suction hopper dredgers pump material from the bed, via one or more suction pipes, into a hopper contained within their structure. At the end of the suction pipe is a draghead, which is designed to maximise the concentration of solids in the pump mixture. Within the hopper some of the solids in the pumped mixture will settle out of suspension while the remaining solids, usually composed of fine sediment, are discharged with the supernatant water, or overflow, via one or more spillways.

The overflow is used mainly when sand is being dredged to allow the hopper to fill with sand and to displace silt-laden water. The overflow is unlikely to operate for more than about 30 minutes (but see aggregate dredging below). When dredging in mud, there is little or no advantage in prolonged overflow because the point is soon reached where the density of sediment being pumped into the hopper is little higher than that overflowing it. Sometimes use of the overflow in mud is restricted because of the risk of environmental impact, especially if the sediment is likely to be contaminated. If overflow is used with mud, it constitutes hydrodynamic dredging as discussed in Section 3.2.11.

Spillways may be weir like structures, as commonly found in small or medium-sized UK aggregate dredging vessels, or a single or double central spillway that discharges under the keel, as is common in capital/maintenance dredging contractor vessels.

Figure 3.1 *Trailing suction hopper dredger used for aggregate dredging in the UK*

The loss of sediment from a trailing suction hopper dredging is principally a result of the amount of overflowing that takes place. This is controlled by the dredge operator, who will make the decision on the basis of obtaining the best economic load but with due regard to any environmental constraints.

Some modern trailing suction hopper dredgers have a "light material over board" (LMOB) system that detects when the material density is less than a pre-set value and diverts the pumped mixture overboard if this is the case. It is not used when dredging only sand and need not be used when dredging only mud. The system helps to maximise the ratio of solids to water within the hopper when dredging a mixture of sand and mud.

Aggregate dredging

In the UK, the aggregate dredging industry operates its own vessels, which tend to discharge overflow via several shipside spillways. Aggregate dredging usually involves extensive overflowing and may additionally require the screening of material, to divide finer fractions from coarser fractions, when a particular sediment fraction is sought. The unwanted sediment fractions are discharged over the side of the vessel. In this case, the discharge of screened material may release much more fine sediment into the water column than the overflow on its own (see Appendix 3 regarding measurements at Great Yarmouth; Section A3.1).

Overflowing, essentially of unwanted fine sediment, occurs during the dredging of sand or gravel where the load can be increased by continued pumping after the hopper has been initially filled with the sediment/water mixture. The proportion of fine material in the sediment is generally relatively small, typically 1–5 per cent, as sites are deliberately chosen for their low silt content. Overflowing is a normal part of the process due to the amount of water that is pumped into the hopper with the desired sediment. When it is initially filled, there may be as little as 10–15 per cent of solids in the hopper. By continuing to pump after the hopper is filled, the amount of solids in the hopper can be increased. The water, containing fine solids that are slow to settle in the hopper, is displaced and overflows, forming a plume in the water as the dredger moves.

Aggregate dredging, however, often entails the screening of dredged material when a particular fraction is required. This practice requires the further discharge of the reject material and results in the increased release of sediment into the water column. A typical aggregate dredging operation will involve trailing in an elliptical path with a major axis length of about 1.5 km. Trailing speeds are usually about 2 knots. A "no screening (all in)" operation may take a little under two hours to fill the hopper, whereas "screening for stone (sand out)" may take up to nearly five hours. Filling the hopper thus takes longer, but in other respects the general description of plume generation by trailing suction hopper dredgers will apply.

Capital and maintenance dredging

Unlike aggregate dredging, contractors usually undertake capital and maintenance dredging with dredgers that overflow via a central spillway, discharging from the ship hull. The dynamic plume produced by discharge from shipside spillways, which are further from the seabed than a central spillway, has a greater initial dispersion in the water. Central spillways on the other hand allow less initial dispersion of the dynamic plume and more of the fine sediment impacts on the bed.

Causes of sediment release

The main causes of sediment release during trailing suction hopper dredging are:

- draghead disturbance at the seabed
- discharge of overflow water via spillways
- discharge of screened material
- LMOB discharge
- turbulence caused by the dredger propeller scouring the seabed
- disturbance of gas in the sediment may enhance re-suspension.

Control

The re-suspension caused by trailing suction hopper dredging can be reduced by optimising trailing velocity, suction position and pump discharge, reducing water intake/overflowing and through the return flow method. Further details are provided in Section 6.1.1.

3.2.3 Suction dredging

Aggregate dredging

Suction dredgers are stationary. They share some of the characteristics of the trailing suction hopper dredger but do not dredge while under way. Suction dredgers anchor and then load the hopper while moored, usually creating a pit in the seabed. Internationally, suction dredgers generally load barges or pump straight ashore, being used for dredging small rivers. In the UK suction dredgers often have their own hoppers and are used for aggregate dredging. These suction dredgers with hopper capacity range from 150 m^3 to 11 000 m^3.

Apart from the effects of a moving draghead and propeller disturbance, the potential for losses are similar to those of the trailing suction hopper dredger.

3.2.4 Cutter suction dredging

Capital and maintenance dredging

Cutter suction dredging (as illustrated in Figure 3.2) involves the cutting or breaking of seabed material by a powerful cutter mounted at the end of a suction pipe, and the pumping of dislodged material by means of a centrifugal pump. Cutter suction dredgers are usually rated by installed horsepower. They range from less than 100 hp to more than 30 000 hp. The dredger commonly takes the form of a pontoon hull structure, but larger dredgers may be ship-form and equipped with propulsion. Material is usually transferred to the point of disposal via pipelines, but it may also be loaded into barges moored alongside the dredger. This method of dredging can be used to dredge a variety of materials, including rock.

A typical operation involves positioning the dredger at the cutting face then swinging the dredger by the use of side wires and cutting a circular segment, gradually moving forward. Apart from the need to shift anchors and pipelines intermittently to enable the dredger to continue moving forward, the production is more or less continuous.

Figure 3.2 *Cutter suction dredger*

Causes of sediment release

The main causes of sediment release during cutter suction dredging are:

- the rotation of the cutter causes centrifugal forces which "throw" material out of reach of the suction and adds to turbulence and re-suspension
- when excavation production exceeds pumping capacity excess material is released
- the disturbance of gas in the sediment may enhance re-suspension
- rarely the material may be pumped into a barge, in which case there will be losses due to splashing and overflow
- anchor and wire movement
- pipeline leakage.

Control

The re-suspension caused by cutter suction dredging can be reduced by optimising the cutter speed, swing velocity and suction discharge, shielding the cutter head or suction head and optimising the design of the cutter head. Section 6.1.2 provides more details.

3.2.5 Dustpan dredger

Capital and maintenance dredging

Dustpan dredgers are similar to cutter suction dredgers but do not employ a rotating cutter head. Instead they use a wide horizontal suction head, which is fitted with an array of water jets to loosen material. Like the cutter suction dredgers they discharge through a floating pipeline. This type of dredging is only used with soft material, being suited to the removal of thin deposits from a large area. They are commonly used in the UK to remove fresh deposits from capital works before placement of pipelines and tunnel elements.

Causes of sediment release

The releases of sediment into the water column arising from this type of dredger are as follows:

- disturbance of the bed due to the action of water jets
- gas released from the bed material putting sediment into suspension
- anchor and wire movement
- pipeline leakage
- splashing and overflow resulting from loading to barges.

3.2.6 Grab dredging

Grab dredging (as illustrated in Figure 3.3) consists of the lowering and raising a grab by a crane. There are many types of grab, some specially designed to minimise losses. The grab capacity generally ranges from less than 1 m^3 to 20 m^3.

The crane may be pontoon-mounted, discharging into a barge, or vessel-mounted, discharging into a hopper contained within the vessel. Pontoon grab dredging is well suited to the dredging of confined areas, as in docks and quays and can be used for a wide range of depths, although the production rates are relatively low compared with other methods. Grab hopper dredging takes place with the vessel anchored. Hopper sizes are generally relatively small (less than 1500 m^3).

Cycle times for filling and discharging the grab depend on water depth, soil type and the operator's skill, but typically are in the range of less than one minute to about two minutes. Interruptions may occur to move the dredger or when the hopper or barge is fully loaded.

Figure 3.3 *Grab dredging*

Causes of sediment release

The main causes of sediment release during grab dredging are:

- the impact of the grab on the bed
- disturbance of the bed during closing and initial removal from bed
- material spilled from the grab during hoisting
- material washed from the outer surface of the grab during hoisting
- leakage and dripping during slewing
- aerosol formation during re-entry
- washing of residual adhering material during lowering
- disturbance of gas in the sediment, which may enhance re-suspension
- spill/splash overwashing of barges.

Control

The re-suspension caused by grab dredging can be reduced by using a watertight grab, using a hydraulic grab, using a silt screen, limiting the swing of the grab over open water, and avoiding levelling the irregular dredged surface by dragging the grab over the bottom. Further details are presented in Section 6.1.3.

3.2.7 Backhoe dredging

Capital and maintenance dredging

Backhoe dredging is analogous to the common land-based excavator mounted on a tracked vehicle. In dredging the excavator is usually mounted at one end of a pontoon. The excavating action of the backhoe requires the pontoon to be fixed in position by spikes or "spuds" pushed into the seabed. The size of bucket used varies with the nature of the material to be dredged and the maximum dredging depth. The method can be used to dredge a variety of materials. It can work in confined spaces and has a quicker cycle time than an equivalent-sized grab dredger. However, the excavator becomes weaker at tearing out material as depths increase and the production rate is less than that for dredgers for which the dredging process is relatively continuous (eg cutter suction and bucket ladder dredgers) (Bray *et al*, 1997). Bucket sizes are generally in the same range as those of grabs. A typical operation is similar to that of the grab, with cycle times dependent on the water depth, the skill of the operator and the difficulty of extracting the material (noting that the plant is frequently selected specifically to deal with a difficult material).

Causes of sediment release

The main causes of sediment release during backhoe dredging are similar to those of grab dredging:

- impact of the bucket on the bed
- disturbance of the bed during initial removal of bucket
- material spilled from the bucket during hoisting
- material washed from the outer surface of the bucket during hoisting
- leakage and dripping during slewing
- washing of residual adhering material during lowering
- disturbance of gas in the sediment, which may enhance re-suspension.
- washing of residual adhering material during lowering
- spilling, splashing, overflow of barges.

Control

The re-suspension caused by backhoe dredging is greatly dependent on operator skill and can be reduced by use of experienced operators. Reduction can also be achieved by use of a silt screen or a visor grab (similar to a closed grab for grab dredging). Further details are presented in Section 6.1.4.

3.2.8 Dipper dredging

Dipper dredgers are similar in principle to backhoe dredgers except that the grab is mounted on different arm geometry and cuts forwards into the dredging face. The range of bucket sizes is similar and the losses and mitigation are similar to those of the backhoe dredger.

3.2.9 Bucket ladder dredging

Capital and maintenance dredging

The dredging action in bucket ladder dredging is achieved by a continuous chain of buckets that scoop material from the seabed and raise it above water. The buckets are inverted as they pass over the top of the ladder and discharge their load onto chutes that convey the dredged material to barges alongside. They are normally classified by bucket size, which ranges from very small up to a little less than 1 m^3. The production is only interrupted when it is necessary to change barges or relocate the position of the dredger.

Causes of sediment release

The main causes of sediment release during bucket ladder dredging are:

- disturbance around the buckets as they are digging
- dragging the bucket chain over the bed when working in shallow water
- sediment spilling and being washed from the buckets as they ascend
- leakage from the discharge chutes
- residual adhering sediment being washed from the buckets as they ascend under water
- release of air trapped in the descending buckets when they turn at the bottom tumbler
- the disturbance of gas in the sediment may enhance re-suspension.
- washing of residual adhering material during lowering
- spilling, splashing, overflow of barges.

Control

The re-suspension caused by bucket dredging can be reduced by optimising the degree of filling of the buckets, adjusting the amount of slack in the bucket chain, controlling the swing, the advance of the dredger, the rate of bucket filling and the bank height, and by good maintenance of the discharge chutes, installing splash screens, inserting one-way valves in the bottom of the buckets and by enclosing the ascending part of the bucket ladder. Further details are presented in Section 6.1.5.

3.2.10 Bed-leveller dredging

Capital and maintenance dredging

This type of dredging involves the mechanical agitation of the bed by a vessel towing a specially designed unit along the bed (Figure 3.4). It is a technique often used in conjunction with maintenance and capital dredging operations in situations where material needs to be moved to a location that is more accessible to a larger dredging plant, or for levelling the seabed after dredging by a larger plant. Re-suspension is inevitable with bed-leveller dredging due to the disturbance of the bed caused by the levelling unit itself and the manoeuvring of the vessel (HR Wallingford, 1996a).

Figure 3.4 *Bed leveller dredging*

3.2.11 Hydrodynamic dredging

Capital and maintenance dredging

Hydrodynamic dredging is a new term that describes all dredging techniques that use the re-suspension of sediments and the transport of the material by means of natural hydrodynamic processes as part of the dredging process. The term covers mixing to make a density current (water injection), stirring into suspension (rakes, harrows and water erosion techniques) and the release of material into the surface waters (hopper overflow, sidecasting, etc) (Van Raalte and Bray, 1999).

The method is used for both maintenance and capital dredging and for both large and small dredging operations. The method is only applicable to some situations but, where feasible, allows a low cost and flexible solution. The technique can be utilised with many types of dredging plant, but there are several types that are specifically designed for hydrodynamic dredging. All create plumes deliberately, as illustrated in Figure 3.5. The principle forms of hydrodynamic dredging are described below.

Water injection dredging

This works on the principle that by injecting water into certain types of material, the *in situ* soil density is reduced to the point where it behaves like a liquid and is mobilised. The method is most successful with low-strength fine-grained material when the opportunity exists for the fluidised sediment to flow to a lower level (Bray *et al*, 1997). The water injection dredger has a pump on board the vessel attached to a pipe, which is lowered by winches to the seabed. Along the bottom of the pipe a fixed array of water jet nozzles expel water at a predetermined pressure. The vessel moves slowly ahead, driving the fluidised sediment before it. This technique is only suitable for the removal of silts and fine sands.

Raising
Side casting
Overflowing

Agitation and erosion
Drag head harrowing

Water injection

Figure 3.5 *Plumes formed by hydrodynamic dredging*

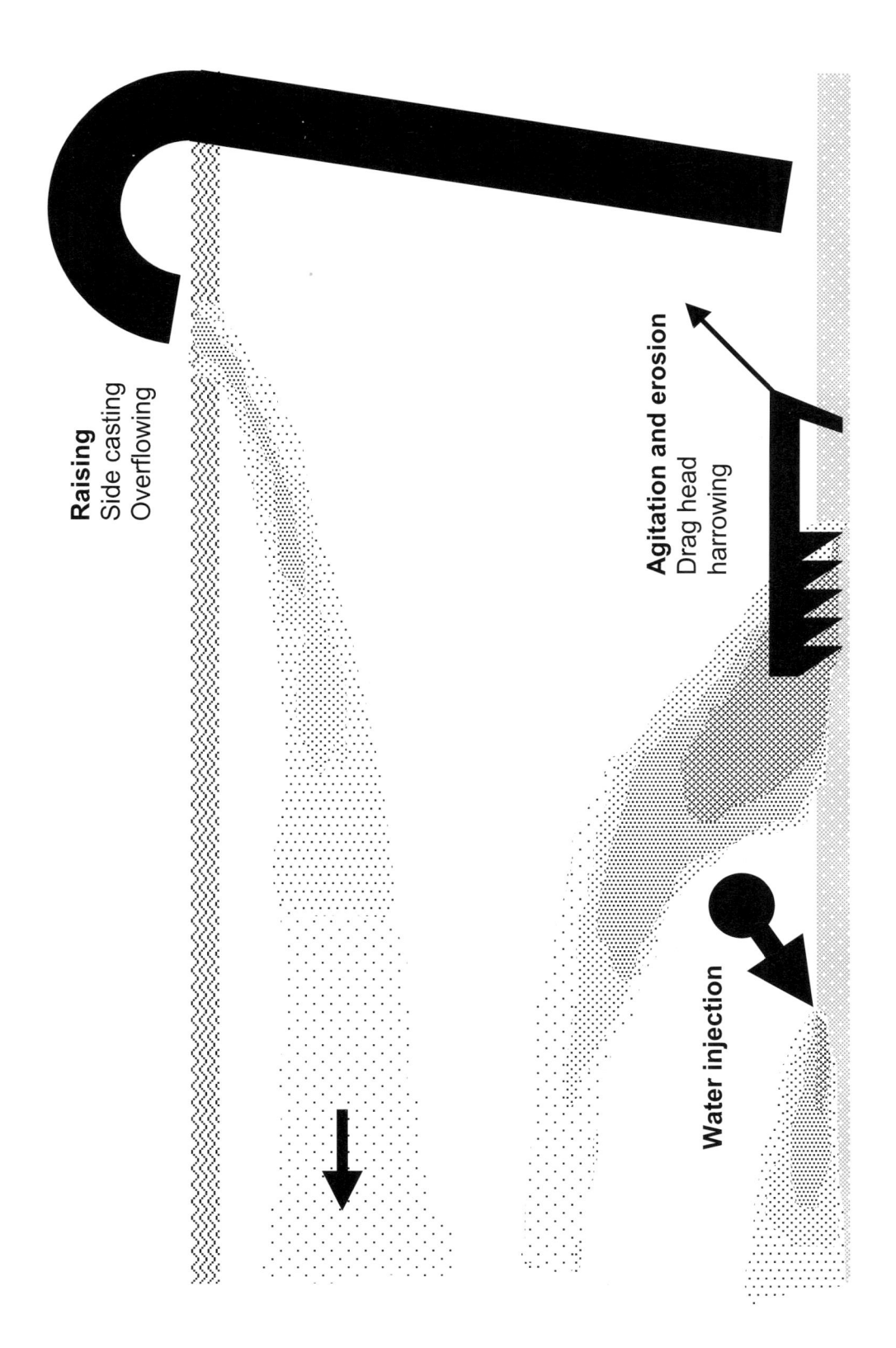

In practice, use of water injection dredging can result in increased suspended sediment higher up in the water column at some distance from the operation (HR Wallingford, 1990). Its use in shallow water or with sediments of high gas content can cause high turbidity throughout the water column (observations at Mistley in the Stour Estuary, 1989, and at Rotterdam, 1986).

Harrowing

In harrowing or raking a unit is towed along the bed to re-suspend material. It only works where currents are sufficient to carry the re-suspended sediment some distance from the dredging location. It has been adopted for both muddy and sandy beds.

Sidecasting/overflowing

Sidecasting is a term given to the technique of deliberately overflowing the hopper or pumping sediment via a fixed pipe from a trailing suction hopper dredger for dispersal by the prevailing flow conditions. Sometimes, sidecasting is via purpose-built spray pipes that throw the sediment/water mixture as far as possible sideways.

Hydrodynamic dredging in the Bristol Channel is illustrated in Figure 3.6.

Figure 3.6 *Hydrodynamic dredging in the Bristol Channel*

3.2.12 Environmental dredging

The term is used for dredging that focuses on operating either with minimal suspension of sediment or with particular accuracy. It can apply to specially adapted variants of any of the types of dredger listed above. There are also some types of dredger and equipment (screens) that have been specifically designed with these objectives in mind.

Auger dredgers

This type of cutter suction dredger use two horizontal augers instead of a conventional cutting head to move material towards the suction head. Pumping is usually by piston action, which enables high-density material to be pumped. The pumped material is usually put into a barge. Because of the high density and relatively low pumping rates there is little splashing or overflow. This type of dredger can only operate in soft sediments (mud).

Disc bottom cutterhead dredgers

The disc bottom cutterhead dredger is another variant of the cutter suction dredger, this time using a horizontal-mounted disc cutter. It is designed to allow the dredging of thin layers of polluted mud with the smallest possible amount of re-suspension or spillage. Hardly any re-suspension and spill has been recorded during employment of this type of dredger (HR Wallingford, 1996a).

Scoop/sweep dredgers

Scoop/sweep dredgers use a specially designed cutting head that scrapes material towards a suction intake. Monitoring of re-suspension by a scoop dredger was reported by Standaert *et al* (1993). The monitoring showed that suspended sediment increases from this type of dredging were less than 5 mg/l^{-1} within 10 m of the dredger.

Silt screens

In some circumstances use of a silt screen can reduce the release of sediment from some stationary dredgers (ie grab and backhoe dredgers). A silt screen is a curtain of cloth suspended from a floating framework down to the bed; it is permeable to water but not to silt. Further details are presented in Section 6.1.7.

3.2.13 Other plume sources

Transportation via pipeline

Many of the dredgers described above transport dredged material via pipeline to land or to barges for reworking. Pipelines are subject to leakage and occasionally breaks, and are therefore a source of potential sediment plume release.

Loading of barges

The loading of dredged material from dredgers into barges is another source of possible sediment release. This can either be in the form of spillage and splashing from mechanical grabs, as described above, or in situations where a sediment/water mixture is pumped into the barge and overflow becomes an issue (as it does in trailing suction hopper dredging).

Figure 3.7 *Cutter suction dredger* Orion *dredging through chalk and clay, south coast of England*

3.3 MEASUREMENT OF SEDIMENT LOSSES

3.3.1 Introduction

In the previous section the potential sources of sediment loss were identified for each dredging class and technique. To determine the relative significance of the sources it is necessary to rely on measurements, yet these are fraught with difficulties because of the dynamic and transient character of the phenomena and the adverse operational situation. In this section, the methods of measurement and the way in which they are generally interpreted is discussed. Examples of the results obtained are given, but the reader is advised to heed the following warning.

A WORD OF WARNING

Although the current state of knowledge is effectively based on the data presented below and in Appendix 3, a review of the field results (HR Wallingford, 1999b) has revealed that **the general applicability of the reported data is limited** for the following reasons:

- the data have been collected using different and incomparable approaches
- measurements of suspended sediment increases have often been measured at different times and at different distances from the dredger
- the data have often been collected without recording important information such as the characteristics of the dredged material.

The variability of these data means that care should be taken in using it to draw conclusions about the relative effects of different types of dredging activity.

The data have been used in this report to illustrate the current state of knowledge. The conclusions are subject to considerable uncertainty and require more widespread detailed investigations to draw conclusions about particular dredger activities or methods of dredging.

Both fine and coarse sediment can be lost. Coarse sediment disturbed from the bed, or lost during other phases of the dredging operation, rapidly falls back to the seabed close to the point of dredging and so does not become part of a passive plume. The silt or fine sand fractions stay in suspension longer and become the constituent of passive plumes, although it is only the silt fraction that contributes significantly to turbidity.

Sediment losses can be described in the following ways:

- as sediment concentration increases in the vicinity of the dredging (mg/l)
- the rate of release of sediment into the water column per unit time (kg/s)
- via the "S-factor" approach – in which the total mass of sediment put into suspension, is expressed relative to the quantity of material that is dredged (kg/m^{-3})
- via the sediment flux method which describes the sediment lost through the boundaries of a designated area within which the dredger is working.

Each of these approaches is discussed in turn below. It should be noted that each provides some evaluation of the loss of sediment during production, but none records the intermittent nature of release over time that characterises real dredging operations.

3.3.2 Suspended sediment concentration increases

Bray *et al* (1997) cites the measurement of suspended sediment increases in the vicinity of the dredger as one of the possible approaches to evaluating the loss of sediment arising from dredging. However, concentration increases are highly site-specific and do not provide a basis for comparison of different dredging methods, operations, hydrodynamic conditions and material.

Table 3.2 summarises available measurements of increases in suspended sediment concentrations arising from the different dredging activities listed in Appendix 3. The table shows that observed increases vary greatly, even for the same type of operation, which reflects the effect of differing measurement approaches and of different site conditions. Although great care must be used when applying these measurements, taken as a whole they appear to imply that trailing suction hopper dredging produces the largest increases in concentration of suspended sediment (although impacts can vary widely). Grab dredging and bucket ladder dredging appear to produce the same order of increases, which, though varying widely, appear to be smaller than trailing suction hopper dredging. Cutter suction dredging appears to produce concentration increases significantly below these dredging methods and the measurements from the environmental scoop method are the smallest of all.

3.3.3 Rate of release per unit time

The rate of release of sediment per unit time (kg/s) is the sediment loss parameter often used as the source term in plume modelling.

In the case of trailing suction hopper dredging, this is readily measured by taken samples from the hopper near the overflow spillways and multiplying the concentration by the discharge (assumed to be equal to the pumping rate). Such dredger-based measurements reflect the start of the dynamic plume phase and, therefore, do not take into consideration the rapid settling of sediment that occurs during the dynamic plume phase. It therefore produces higher estimates than when using measurements of the suspended solids concentration in the surrounding waters.

The reduced source strength can be estimated from such measurements using existing mathematical models but, in general, such models have not been well validated, owing to the absence of reliable field data. In the case of other methods, the estimation of sediment loss rates in kg/s again involves field monitoring of concentrations around the dredger, although mathematical models of releases exist for some dredging methods (see Appendix 3). Loss rates in kg/s for applications are shown in Table 3.2. Again, even allowing for differences between the dredger-based and water-based measurements, it can be seen that the losses are largest for trailing suction hopper dredging, with capital dredging in Hong Kong showing the highest levels. This was a sand-winning exercise that involved deliberate and significant use of overflowing. Losses for the other methods vary but seem to largely be of a similar order of magnitude to each other, except where environmental measures have been used.

Table 3.2 *Example values for the mass of sediment re-suspended and lost during dredging (mg/l and kg/s)*

A cautionary approach should be taken to the use of these data (see warning above)

Dredging operation	Suspended sediment concentration increases (mg/l)	Fine sediment losses (kg/s)
TSHD (Owers bank, UK)	29–209 (silt)	
	611–2117 (sand)	
TSHD (Great Yarmouth, UK)		20
TSHD (Hastings, UK)		14
TSHD (Hong Kong)		280*
TSHD (Rotterdam, the Netherlands)	150–450	
TSHD (Delfzijl)	10–20	
TSHD (Grays Harbor)	Up to 700	
TSHD (Mare Island)	Up to 1100	
TSHD (Richmond Harbor)	20–200	
TSHD (Alameda Naval Air Station)	40–190	
Grab (Merwdehaven)	35	
Grab (Hollandsche IJssel)	2–100	
Grab (Zierikzee)	90–105	
Grab (Black Rock)	720–1100	1.68
Grab (Duwamish Waterway)	70–160	
Grab (Calumet River)	30–130	0.24
Grab (Alameda Naval Air Station)	29–214	
Grab (Ketelmeer)	up to 5 (depth-avg)	
CSD (Land in Mott McDonald, 1991)		1.33
CSD (Hong Kong)		0.9–1.6
CSD (Calumet Harbur)	2–5	
CSD (James River)	42–86	
CSD (Ketelmeer)	7 (depth-avg)	
Bucket ladder (Barnard, 1978)	Up to 500+, 100 average	
Bucket ladder (Noordzeekanaal, Rotterdam)	110	
Bucket ladder (SATURN)		4
Bucket ladder (Aalschover)		0.33
Bucket ladder (Ketelmeer)	30 (depth-avg)	
Scoop (Standaert *et al*, 1993)	2–5	

* The magnitude of this loss figure reflects the type of dredged material and operation and is not considered to be exceptional under these circumstances.

3.3.4 Rate of release per unit of dredged material ("S-factor")

The rate of release per unit of dredged material is an "average" loss quantity that can take into account differences between dredging methods and enables comparison between methods in terms of total mass lost into the surrounding water. This can be of primary importance when considering contaminated material.

Blokland (1988) describes a systematic method for determining the quantity of sediment released into the immediate vicinity of the dredger, in kg/m^3 dredged. This is the so-called "S-factor". It is based on *in situ* measurements of suspended sediment concentrations, rather than measured release rates from the dredger and so takes into account any dynamic plume effects. The S-factor can be a useful first estimate of the magnitude of sediment losses for situations where there is no site-specific information.

The S-factor measurement is based on the measurement of the suspended sediment concentration distribution of the entire plume during steady state dredging. This information is used together with the (measured or estimated) settling velocity to produce an estimate of release rates. Both measurements can be problematic and open to uncertainty. The S-factor is meant to be representative for both the source term and the dynamic plume.

Kirby and Land (1991) made further use of the S-factor, and the table from their paper is reproduced in Table 3.3. The table is a summary of the sorts of losses of fine sediment that result from the different types of dredging operation in muddy sediment. It agrees broadly with the summary of field study data presented in Appendix 3, although it does not include results for the prolonged extensive overflow and/or screening discharges that can often occur during aggregate trailing suction hopper dredging.

Table 3.3 *Indicative values for the mass of sediment re-suspended per m^3 of dredged material (Kirby and Land, 1991)*

A cautionary approach should be taken to the use of these data (see warning above)

Dredger type	S-factor (kgm^{-3})
TSHD (no overflow)	typically 7
TSHD (no overflow or LMOB)	typically 3–4
Grab (open, no silt screen)	12–25*
Grab (closed, no silt screen)	11–20*
Grab (closed, with silt screen)	2–5*
Bucket ladder	15–30*
CSD	approximately 6
CSD (reduced swing and rotation speeds)	approximately 3
Dustpan	approximately 4
Backhoe (no silt screen)	12–25*
Backhoe (with silt screen)	5–10*
Auger**	5
Auger (reduced rate of advance)**	3

*   depending on the size of grab or bucket – the smaller the grab/bucket, the greater the re-suspension
**   these auger dredgers were not designed as environmental dredgers

Table 3.3 shows that the losses resulting from the digging-type methods (grab, bucket and backhoe dredging) are significantly higher than those for the suction-type methods (TSHD and CSD). Cutter suction dredging appears to cause the least losses overall, together with the more environmentally focused dredging methods such as dustpan and auger dredging.

The S-factor provides a measure of the total amount of sediment loss during dredging, which is independent of the production rate. The methods with faster production rates (eg trailing suction hopper) actually produce greater increases in suspended sediment concentrations. When concern is focused on the magnitude of suspended sediment increases resulting from different types of dredging, trailing suction hopper dredging is often found to have the largest impact (see Table 3.2). However, the S-factor measurements indicate that trailing suction hopper dredging can result in lower total losses of dredged material over a total project than grab, bucket and backhoe dredging.

3.3.5 Sediment flux method

The sediment flux method defines sediment losses as "the amount of sediment leaving the work zone". The flux of sediment at the limits of the work zone is found by measuring the current velocities and sediment concentrations across the plume cross-section. This method was chosen for evaluating sediment losses arising from the Øresund Link project (Box 3.1). It can be more relevant in terms of environmental evaluation but it ignores everything that happens between the dredger and the point of measurement.

The results of the Øresund monitoring show that, for broadly the same material type, the cutter suction dredging spillage was of the same order as that from dipper and backhoe dredging. This result differs from those presented in Tables 3.2 and 3.3. For cutter suction dredging, the effect of current speeds and dredging material also influenced the spillage greatly. For dipper and backhoe dredgers the variation in spill with environmental conditions was less significant. The results also show considerable variation in spillage even for the same dredgers working in similar areas in similar conditions.

3.4 RECENT FIELD STUDIES

Field studies have been carried out to investigate the nature of the plumes arising from different types of dredging activity.

Organisations representing Dutch Dredging Contractors and Public Bodies formed the Dredging Research Association (CSB), and commissioned Delft Hydraulics to carry out the research that has produced some of the most definitive data available (Pennekamp *et al*, various publications). It was this work that resulted in the definition of an S-factor, that is a measure of the amount of solids released into suspension per unit volume of material dredged. A list of the field studies and a summary of the results obtained is given in Appendix 4.

Box 3.1   *Use of the sediment flux method during the Øresund Link project*

> Monitoring during the dredging activities undertaken for the Øresund Link project (described more fully in Appendix 4, Section A4.9) defined the work zone as the area within 200 m of dredging. The sediment lost from this area was calculated as the product of suspended sediment concentrations and current speeds through the work area boundaries. The sediment fluxes from the dredging activities were then expressed as a percentage of the production rate, defined as spillage, and were monitored closely to make sure that environmental limitations set by the project were maintained (Figure 3.8). This method thus presents an alternative to the S-factor approach but relies on accurate and detailed monitoring of current speeds and plume concentrations.
>
> The monitoring carried out as part of the Øresund Link project represents a very valuable data set due to the well-described and constant approach to measurement. Monitoring was carried for cutter suction dredging of limestone, moraine clay and sand and for dipper and backhoe dredging of moraine clay.
>
> Lorenz (1999) gives the following summaries of spillage from this project.
>
> **Table 3.4**   *Summary of the spillage from cutter suction dredging*
>
> | Dredging operation | Current speeds | Material type | Spillage % |
> |---|---|---|---|
> | CSD | >0.5 m.s^{-1} | limestone | 6–8 |
> | CSD | >0.5 m.s^{-1} | clay till | 3–5 |
> | CSD | <0.5 m.s^{-1} | limestone | 3–6 |
> | CSD | <0.5 m.s^{-1} | clay till | 2–4 |
>
> **Table 3.5**   *Summary of the spillage from dipper and backhoe dredging*
>
> | Dredging operation | Location | Material type | Spillage % |
> |---|---|---|---|
> | Dipper and backhoe | large open areas | clay till | 1–4 |
> | Dipper and backhoe | narrow channels | clay till | 2–5 |
> | Dipper and backhoe | bridge pier foundation pits | clay till | 4–6 |

More recent plume studies include those associated with the following research initiatives and dredging projects:

- Plume Measurement System of the USACE Dredging Research Programme
- channel deepening to Londonderry, Lough Foyle, Northern Ireland
- agitation dredging at Sheerness, eastern England
- construction of pipeline trench, southern England
- maintenance dredging on the River Tees, north-eastern England
- aggregate dredging operations in the UK including the Marine Aggregates Mining Benthic and Surface Plume Study funded by the US Minerals Management Service, and monitoring of aggregate dredging plumes at Race Bank by CEFAS
- dredging activities in Hong Kong (various)
- monitoring associated with the Øresund Link, Denmark and Sweden.

These studies are described in Appendix 4 and the key findings are summarised below.

The use of acoustic profilers (see Glossary for definition) in combination with other measurement systems and water sampling has led to a considerable increase in the understanding of plume processes. A description of this equipment is given in Land and Bray (1998). In particular, measurements have consistently noted a rapid decay in suspended sediment concentration increases with distance from the dredger. Using acoustic profiler backscatter, plumes can be successfully tracked for several kilometres at concentrations close to background levels. The recent monitoring undertaken for the Øresund Link has shown that a combination of measurement systems can be used successfully to monitor, and manage in real time, the losses arising from dredging activities, although at a very high cost.

Minipod measurements suggest that under certain conditions, near-bed concentration increases of 50–150 mg/l above background conditions can result from the advection of passive plumes induced by aggregate dredging, even 6–7 km from the point of dredging.

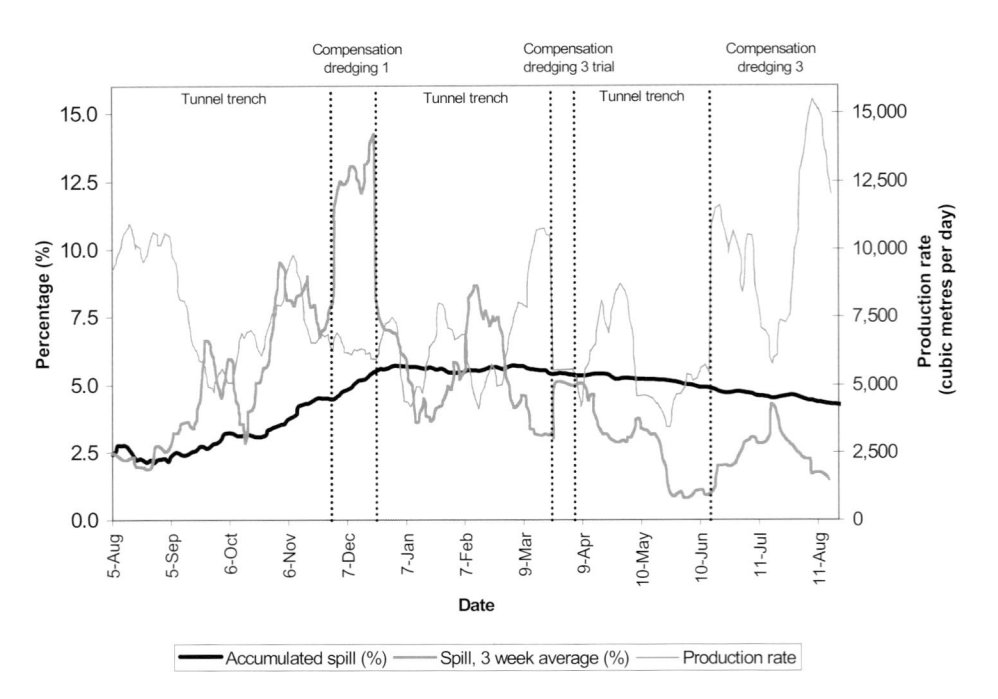

Figure 3.8   *Changes in spill and production by cutter suction dredger* Castor *1996–1997 (from Lorenz, 1999)*

On-board sampling from trailing suction hopper dredgers undertaking aggregate dredging has enabled a comparison of the amount of fine sediment released from the dredger with the amount of fine sediment initially released into the water column as a passive plume. The results of these studies have confirmed that processes occurring during the first few minutes of the plume generation are responsible for the loss to the bed of a considerable proportion of the fine sediment initially released into the water column. These studies have also enabled the quantification of the proportions of different fractions discharged into the water column from these activities.

Field studies undertaken in Hong Kong have indicated that the most important factors influencing this dynamic phase appear to be the initial momentum and increased density of the plume with respect to the surrounding water, although there is evidence that the aggregation of fine, muddy material to coarser grains is also important.

4 Plume processes and modelling

Chapter 3 considered the sources of sediment plumes. This chapter focuses on the behaviour of plumes or plume processes. Available research on the dynamics and modelling of sediment plumes is reviewed with a view to understanding how they may affect the aquatic environment.

The main physical processes affecting the advection and dispersion of sediment plumes have been identified and the state of knowledge regarding these processes has been addressed. They are summarised in Table 4.1.

Table 4.1   Physical processes affecting sediment plume dispersion

Physical process	Cause
Advection	Primarily currents, but these can be affected by waves and wind
Dispersion/ diffusion	Variations in currents and waves Turbulence
Settling on bed	Rapid descent of dynamic plume Deposition of sediment from passive plume
Release into water column	Turbulent mixing from rapid descent of dynamic plume Impact of dynamic plume on bed turbulence caused by propeller scour and vessel movement Erosion from bed deposit arising from descent of dynamic plume re-suspension of sediment deposited from passive plume

This chapter does not consider the movement of coarser sediment fractions that are disturbed by the dredging process. These fractions fall immediately to the bed and are moved as bedload by waves and currents. This type of sediment transport lies outside the remit of this study.

First, the research that has been carried out is reviewed. The processes affecting dynamic plumes and passive plumes are discussed in turn, and the chapter concludes with a review of the state-of-the-art with respect to modelling plumes, details of which are given in Appendix 5 and Appendix 6.

4.1   SUMMARY OF RESEARCH

Most recent research has focused on the dynamic plume phase. The processes involved in the passive plume phase are quite well understood, except for settling velocity.

The work that has been undertaken to date on dynamic plumes has focused on those formed during disposal and then, primarily, on hopper discharge through bottom opening. Theoretical studies have been undertaken by Koh and Chang (1973) for the US WES, which assume that dumped material behaves as a fluid and which may be appropriate for the dynamic stages of overflow plume development.

Another approach has been studied by Krishnappen in Canada (HR Wallingford, 1999b) in which the material as treated as discrete soil particles. This approach is not likely to be appropriate for most overflow operations but may be suitable when considering very coarse releases during certain types of aggregate dredging. Ogawa in Japan has also undertaken laboratory studies suited to coarse material, but which did not include the effects of entrainment and dilution (HR Wallingford, 1999b).

A large amount of research has been undertaken in the UK and in the USA to characterise the operational effects of dredging on the release of fine sediment into the water column. This has been with the aim of providing mathematical models that will enable the prediction of losses in a form that can be used as input to plume prediction studies. These models have built upon the considerable amount of field information that has been obtained over the last decade (summarised in Appendix 3). Computer models have been created that will predict the release of sediment caused by grab, cutter suction, bucket ladder, backhoe and trailing suction hopper dredgers, but they lack calibration data (HR Wallingford, 1999b). However, this sort of work has contributed to the understanding of source terms of sediment release from dredging activity which in turn has led to a better understanding of the dynamic plume phase.

In the Netherlands, Delft University is undertaking research on the modelling of overflow, especially the behaviour of the dynamic plume. Laboratory experiments are carried out to determine the circumstances under which overflow results in a dynamic plume.

In recent years, the losses of sediment associated with trailing suction hopper dredging for aggregates has been the focus for research targeted towards a better understanding of the key processes affecting the behaviour of dredging plumes. The field and modelling work has led to the knowledge that the behaviour of the dynamic plume is of primary importance in determining the magnitude of turbidity increases as a result of dredging.

In Hong Kong, studies have been directed at establishing the losses during dredging and the subsequent advection and dispersion of the plumes of fine sediment so generated. Field measurements using the acoustic profiler techniques have investigated the processes occurring during the first few minutes of the plume generation (Whiteside *et al*, 1995). These showed that the initial processes are likely to be responsible for the loss to the bed of a considerable proportion of the fine sediment initially released into the water column.

HR Wallingford (1996b) also investigated the processes that may be occurring during this initial phase of the development of a plume. It was concluded that, for the vessels operating in Hong Kong with a single sub-surface spillway, the initial momentum of the discharge from the vessel is a significant factor, resulting in much of the sediment descending directly to the seabed. Additionally it was postulated that the disaggregation of fine muddy material during the dredging process is not complete and that a further significant proportion of the muddy material released into the water column may be in the form of fine clay balls, or adhered to coarser grains. Thus some of the re-suspended fine material may settle to the bed out of the passive phase of the plume with settling velocities in excess of that of a natural muddy suspension.

4.2 PROCESSES AFFECTING DYNAMIC PLUMES

The factors affecting dynamic plumes have been identified as follows:

- the nature of the source material
- the density and momentum of the descent phase
- aggregation of particles in the descent phase
- various factors affecting horizontal bed movement phase.

These are discussed in more detail in the following sections.

4.2.1 The nature of the source material

The material forming the source of a dynamic plume is a water/sediment mixture. Its properties are different from those of the *in situ* material because of the dredging process. The main effects of dredging are the addition of water and disaggregation due to mechanical and hydraulic forces.

Research undertaken by VBKO (HR Wallingford 1999b) has identified the need for a simple standard test to determine the disaggregation index for a material to be dredged. Traditional soil mechanics tests do not give a reliable measure although such parameters as liquidity and plastic indices give some indication. Tests resembling abrasion tests seem to offer the most promising direction.

4.2.2 Density and momentum in the descent phase

The descent of the dynamic plume is so relatively rapid that horizontal advection processes are virtually negligible. The process of discharging sediment into the water column (ie with an initial momentum), together with the negative buoyancy effect caused by the greater density of the sediment plume, gives a high downward velocity and momentum to the plume.

As the plume descends, dilution occurs as water is entrained. It has been hypothesised that the entrainment of ambient water into the plume is proportional to the local velocity of the plume and to the surface area of the plume/ambient water interface.

It is thought that dilution does not affect the dense core region. The outer region, however, may be considered to be a passive plume. The boundary between the two has no precise definition but dynamic plume models assume a value of suspended solids concentration of 250 mg/l (HR Wallingford, 1999b).

A series of sensitivity tests were undertaken (HR Wallingford, 1999a) using a numerical model of the initial phase of plume dispersal developed from USEPA effluent outfall plume models (USEPA, 1985). These are designed to predict the average dilution of the buoyant plume as it ascends. The rate at which mixing occurs between these two bodies of water was assumed to be proportional to the relative velocity between the plume and the ambient water and the surface area of the plume. The computational modelling tests showed that, with the possible exception of larger aggregates such as clay lumps, the effects of initial momentum and negative buoyancy are the most significant in determining the initial dispersion.

The speed at which plumes descend under the influences of these factors depended on the type of dredging operation in question. Speeds were, in some cases, orders of magnitude higher than the settling velocity of individual sediment particle grains.

Further experiments showed that the amount of sediment re-suspended is more dependent on the depth of water and whether central hull or shipside discharge is used. The discharge rate was a secondary factor. The greater the depth of water, the more opportunity there was for sediment to be stripped from the plume into the ambient waters to form a passive plume. However, no representation of re-suspension due to the impact of the dynamic plume on the bed was included in the models used.

Further research was undertaken in the UK using a prototype dynamic plume model (HR Wallingford, 1999b). This model reproduced the rapid descent of the dynamic plume and included a criterion involving suspended sediment concentration for deciding the amount of fine sediment released into the water column. Again the re-suspension caused by the impact of the dynamic plume on the bed was not included in the model.

The modelling work, again, highlighted the importance of the height of the overflow (or screening discharge) release above the seabed, but suggested that the initial discharge or momentum of release had little or no effect on the proportion of sediment entrained into the ambient water. The work highlighted the need for a better understanding of the mechanism for mixing between the dynamic plume and the surrounding water, in particular for estimating total losses of sediment to the water column.

Other work (Hydroynamic, pers comm) on the physical fate of sidecasted material has shown that the dynamic plume is largely determined by the initial vertical velocity, plume diameter and density.

4.2.3 Aggregation during descent

The degree of aggregation of the sediment affects the settling velocity of sediment in the dynamic phase. Research undertaken for the DETR regarding dynamic plumes caused by trailing suction hopper dredgers (HR Wallingford, 1999a) investigated the relative importance, during the initial moments of release, of the processes of enhanced settling due to aggregation and that due to the momentum and density of the plume.

To investigate the possibility of preferential settling through the aggregation process, the method adopted was to take samples from the spillway of a dredger during aggregate dredging and, in a controlled experiment, observe the settling characteristics of the sediment in the overflow using a specially designed settling column. The nature of the experiment was to some extent artificial, as the settling characteristics of the captured sample are determined in the quiescent environment of the column as opposed to the continuously moving turbulent waters of the sea. This effect could lead to an acceleration of the settlement of fine sediment in the column and an overestimate of the settling velocity.

Another limitation on the experiment was that because the dredged material was poured into the settling column through a sieve (which was used to remove the larger coarse sand and gravel fractions), the initial momentum was reduced. Also, the friction of the settling tube walls affected the plume descent velocity.

Observations of the test showed that the dredged material was reasonably well mixed throughout the top of the water column within a short time. Thus the main process by which any material settled preferentially below an initial mixing depth was due to increased settling velocity caused by aggregation.

It was concluded that the fine sediment in the water column settled faster because of flocculation. The effect of this process was only a small enhancement of settling of fine sediment, although it was recognised that this might not be the case if the experiment had been carried out with larger aggregates such as lumps of clay.

Further research into the effect of aggregation is required.

4.2.4 Flow of the dynamic plume along the bed

The dynamic plume, after impacting the bed, moves radially outwards along the bed as a suspension with sediment concentrations up to 10 000 mg/l. The suspension moves as a density current, the turbulence generated within the plume at first being sufficient to prevent deposition. As the plume moves outwards it loses its initial momentum. If additional energy is imparted, eg by currents, wave energy or bed slope, it may continue to move as a suspension. When the energy level becomes insufficient to maintain it, the suspension can collapse and form a fluid mud layer (HR Wallingford, 1999c).

Fluid mud will flow with the current and with bed slope until the bed shear stress (ie the shear stress between the fluid mud and the bed) falls below 0.1 N/m^2, whereupon it will dewater. Below a bed shear stress of 0.1 N/m^2 fluid mud will dewater to form a weak deposit of dry density in the order of 100–300 kg/m^3 (bulk density 1060–1180 kg/m^3) (HR Wallingford, 1999b). During the fluid mud phase the overlying water will be entrained into the fluid mud. The rate at which this occurs is dependent on the difference in speed between the overlying water and the fluid mud, and the fluid mud concentration, thickness and density.

The state of knowledge of dense concentrations and/or fluid mud is the scope of much research at present and collaborations such as the European COSINUS project are greatly increasing the scientific understanding of these phenomenon. However, as yet there are no well-developed methods for predicting their behaviour due to the large variation in mud properties that can occur. The use of field data is considered to be the most reliable method of assessing their behaviour (HR Wallingford, 1999c).

4.2.5 Re-suspension from the dynamic plume phase

This topic has already been covered in Chapter 3 with respect to the dredging processes that cause dynamic plumes. The main points are summarised here for continuity.

Sediment can be released into suspension to form a passive plume at the following stages:

- during descent as entrainment and dilution take place
- during impact on the bed, depending on the nature of the bed material
- during horizontal movement due to turbulence.

4.3 PROCESSES AFFECTING PASSIVE PLUMES

This section constitutes a review of the state of knowledge on the physical processes affecting plumes. The factors affecting passive plumes are:

- the material itself, primarily its settling velocity
- currents, causing advection and turbulence
- the additional effect of wind and waves.

These are discussed in the following sections. This is followed by a brief discussion of deposition and re-erosion, which, although it is outside of the scope of this report, can form an important part of environmental assessment relating to sediment plumes.

For many of the physical processes, such as the hydrodynamic processes of waves and currents, the state of knowledge is very good and the processes are well described. However, cohesive sediment properties are not well described and there is often some uncertainty in the representation of the behaviour of this material.

4.3.1 Material properties (settling velocity)

The factor that most affects the length of time a particle will stay in suspension is its settling velocity. This in turn affects the distance the plume is able to travel before all of the sediment has settled to the bed and the hydrodynamic conditions themselves affect the settling velocity. Turbulent conditions can maintain very fine sediment particles in suspension indefinitely. However, this rarely happens in nature.

Suspended sediment settling velocity is a parameter that is more important when considering the passive plume than the dynamic plume, as the speed of descent of the dynamic plume is far greater than the settling velocity of individual sediment particles.

The settling velocity of non-cohesive material is well understood and can be specified for a given grain size. The settling velocity of cohesive material varies with the density and size of flocs, which are themselves products of the concentration, turbulent structure and salinity of the water column. It also varies with the density and viscosity of the water. This is particularly important in areas where the water may be stratified by either salinity or temperature.

The relationships between cohesive sediment settling velocity and conditions within the water column are poorly understood. In the past, attempts have been made to find empirical relationships based on laboratory results. However, the problems of reproducing real conditions in the laboratory, in particular in reproducing the flocculation process and turbulent structure, have led more recently to the pre-eminence of *in situ* field measurements.

Such measurements (eg Dyer *et al*, 1996; Feates *et al*, 1999) have shown that median settling velocities can vary considerably (eg –10 mm/s) under different conditions. Recent work by Winterwerp (1999) and Winterwerp *et al* (1999) has constructed an analytical framework for the flocculation, settling and re-suspension of flocs under the action of turbulence. This may allow a greater understanding of the main processes affecting cohesive material in the water column.

In plume dispersion there can be an additional complication. In natural background sediment, particles are constantly interacting with each other with the result that the distribution of particle sizes and density remains approximately constant over time. In the early stages of plume formation there may be little interaction between particles, resulting in the larger, denser particles settling first, leading to a steady reduction in median settling velocity over time.

For most practical applications, the current state of knowledge of cohesive sediment is sufficient for an adequate representation of settling behaviour. In many studies the case of a single value for settling velocity (usually of the order of 1 mm/s^{-1}) has been adopted.

In situations of greater complexity, such as the time-dependent settling alluded to above, the settling behaviour can be represented by characterising the plume behaviour into two or more phases, each with a different settling velocity.

More research is needed to study the actual settling behaviour of cohesive material in passive plumes.

4.3.2 Current flow

Generally, currents are generated by river flow or tidal flow. They may also be induced by density differences, wind and waves. Whatever their origin, their principal effect on passive plumes is advection. While particles remain in suspension they travel with the body of water in which they are situated until they hit the bottom.

In UK waters, tidal currents are usually the main advective and diffusive influence on passive plumes. The main diffusive effect of tidal currents is caused by the changes in velocity (speed and direction) that occur throughout the height of the water column, causing suspended particles to move at differing speeds in different directions, thus causing the spreading of the passive plume. Small-scale fluctuations in currents (turbulence) are another mechanism whereby diffusion occurs. The magnitude of tidal currents also contributes to the erosion and deposition of sediment; this is considered below in Section 4.3.5.

For UK dredging, the other types of currents, (river, density-induced, wind-induced and wave-induced) are generally (but not always) of a lesser magnitude than tidal currents. However, when superimpsed onto oscillatory tidal movement, they can create residual currents. These can influence the net movement of a passive plume.

In the UK, river currents are usually of low importance for most dredging activities, except during storm runoff in estuaries of relatively small tidal range. Elsewhere in the world, in estuaries of large river basins or where there is little tidal variation, these currents can dominate the advection of dredging plumes.

Density currents occur primarily in estuaries and are caused by the gradient in salinity between the freshwater of rivers and the saline water of the open sea. The gradient causes a residual circulation of water upstream near the bed and downstream near the surface. The strength of these currents is greater for estuaries with low tidal ranges and high fluvial flow. In estuaries with a low tidal range, conditions may occur where the surface fresh water is constantly in the ebb direction and oscillatory near the bed. A plume that is settling towards the bed in such circumstances may reverse direction as it falls into the lower layer. Deliberate overflow of a trailing suction hopper dredger was stopped in the Tees Estuary when it was discovered that instead of the currents carrying the sediment out of the estuary they were actually creating a net upriver transport.

Calibrated computational models are capable of making very reliable predictions of current movements. Nevertheless, the practicalities of acquiring, and reproducing, observations of current fields in situations where 3D effects are important are still often regarded as prohibitive in terms of cost and time.

Turbulence is the small-scale temporal and spatial variation of current velocities induced by currents and waves. The effect of turbulence on sediment is to cause mixing (dispersion/diffusion) and to keep particles in suspension. The effects of turbulence on passive plumes are currently reproduced well by random walk or advection/ dispersion models.

4.3.3 Additional effect of waves

Waves are caused by the action of wind over the surface of water, in a process that is well understood and sophisticated models exist for reproducing and forecasting such behaviour. Once formed, waves are subject to the effects of refraction, diffraction, friction, set-up, non-linear interaction with other waves and breaking. All of these processes can be reproduced well by existing computational models to the level of detail needed for plume dispersion studies.

When waves break, they apply a force to the surrounding water, which can form wave-driven currents. Such currents are normally small in magnitude offshore but become significant where wave heights are large, at beaches for example, or where the tidal range is low, and therefore can effect the dispersion and advection of passive plumes.

The passing of a wave causes a fluctuation in current speed near the bed, which adds to the turbulence, and hence dispersion and the potential for erosion. Where wave action alone is present, this process can be described very well. Where tidal currents and waves are both significant the system becomes complex and is a focus for ongoing research.

4.3.4 Additional effect of wind

The action of wind across water, as well as generating waves, can induce wind-driven currents. Since wind-driven currents are usually superimposed on the oscillatory pattern of tidal currents, the effect they produce is that of a residual or "net" current at the water surface in the direction of the wind. The magnitude of wind-driven currents is usually small, but in relatively high winds or areas of low tidal range the wind-driven currents can be significant in influencing the movement of passive plumes. The impact of significant wind-driven currents is different when the wind is blowing to or from shore, or in an enclosed area, to when the wind is blowing alongshore. In the former case, there is a residual surface current in the direction of the wind with an opposite residual current near the bed. In the latter case, there is a residual current, greatest near the surface, in the direction of the wind all through the water column.

The action of wind can be well represented by existing three-dimensional computational models.

4.3.5 Sediment deposition and erosion

Although outside of the scope of this study, the processes of deposition and erosion are briefly discussed below.

The passive plume is advected by tidal currents until at some point, under quiescent conditions, such as at slack water, the deposition of fine sediment will occur. This will reduce the concentrations in the plume and form a deposit on the bed that may or may not be re-eroded, depending on the shear strength of the material itself and the shear stress exerted by the flow. It is an important consideration because it means that sediment may be transported longer distances by a ratchet mechanism than would be possible in a single tidal excursion.

Even if most of the sediment is re-eroded, the time taken for re-erosion causes further dispersion of the original plume and the concentrations of the suspended sediment will be reduced, often to the point of being indistinct from background values. Although not negligible, the re-suspension of sediment deposited from the passive plume is therefore of secondary importance when considering the behaviour of the passive plume, unless background concentrations are particularly low. They may be important when the fate of material in the water column is of importance, as when the dredged material is contaminated and water quality issues are to be considered.

The deposition of sediment is usually represented by most modellers using the standard equation of Krone (1962). This states that the rate of deposition is a function of concentration, settling velocity, bed shear stress and the critical bed shear stress for deposition. This latter parameter varies with the nature of the sediment, but the range of possible values is relatively small and does not usually have a significant effect on rates of deposition.

Erosion of material occurs when the shear force induced by the flow of water on the seabed is sufficient to displace non-cohesive particles or to loosen the cohesive bonds between flocs. The erosion of non-cohesive sediment has been well researched (eg Van Rijn, 1984) and is well understood. The erosion of cohesive sediment is usually represented in computational models by similar equations to that of Owen (HR Wallingford, 1989) or Partheniades (1965), which state that the rate of erosion is a function of the bed shear stress, critical shear stress for erosion and an erosion constant. The latter two parameters are functions of the cohesive sediment. The rate constant can vary by two orders of magnitude (HR Wallingford, 1994), while the critical shear stress for erosion can vary by up to one order of magnitude, depending on whether the sediment is freshly deposited or consolidated. In the case of plume studies it is the erosion of freshly deposited or fluid mud-like cohesive sediment that is of importance so that uncertainty in sediment properties is reduced to values that allow greater erosion.

A greater understanding of representative values of the critical shear stress for the erosion of freshly deposited sediment would aid the modelling of passive plumes. Also, representative parameters are needed for the erosion of sediment from the dynamic plume after impact with the bed.

4.4 MODELLING TECHNIQUES

In assessing sediment plumes arising from dredging it is clearly important to be able to predict both the plume behaviour and its impact rather than simply to measure it afterwards. This implies a need for well established and validated models. Some parts of the processes can be simulated to a more than adequate degree, but others cannot. This section provides a brief summary of various modelling approaches used for different aspects of dredging plume prediction. A fuller discussion is included in Appendix 5. A comprehensive review of techniques is outside the scope of this report, but the existing state of knowledge is described for the purposes of identifying future research needs (see Sections 7 and 8).

Modelling falls into the following broad categories:

- flow models
- dredging process models
- dynamic plume models
- fluid mud flow models
- passive plume models
- water quality models.

It is not considered necessary to discuss the details of flow modelling in this report because it is sufficiently well established and well known. Nevertheless, the choice of 2D or 3D modelling, random walk or advection/diffusion modelling, finite element or finite difference and the type of grid, regular or variable, is an important one, as it will be the driver for passive plume modelling, water quality modelling and possibly for any subsequent environmental modelling.

Models have been developed by several hydraulic laboratories that would be willing to supply further details. It is not appropriate here to indicate any author preference. The review therefore focuses on the state-of-the-art of the various techniques.

4.4.1 Dredging process models

These define the source term that will be used as an input to either the dynamic plume model or the passive plume model. Various models exist, primarily in the USA. Most are empirical, being based directly on field measurement rather than representing the whole dredging process with appropriate algorithms. Work is in progress in Europe on the development of such models for a limited range of types of dredger (including those most commonly used). The models exist in pilot form but require calibration data in order to establish appropriate coefficients. The process models need to include a soil model that describes the behaviour of the material under the action of dredging, in particular its disaggregation.

4.4.2 Dynamic plume modelling

These have to represent the descent, bed impact and horizontal spread of a fluid sediment mixture. Models have been developed in the USA, Canada and Japan and recently in the UK. Methods range from assuming the plume is made up of separate particles that behave independently to those that assume the plume behaves as a dense fluid. Those based on the former assumption are unlikely to be of much use in the context of plume impact assessment.

Under the Dredging Research Program of the US Army Corps of Engineers such a dynamic plume model has been included in a numerical PC tool called STFATE, for assessing the short-term fate of material in open water (USAWES, 1995). This STFATE model includes both the dynamic phase as well as the first hours of the passive plume phase as defined here.

4.4.3 Fluid mud flow modelling

This type of modelling has to represent the horizontal movement of a high-concentration plume on the bed. It is applicable to the last phase of the dynamic plume model. It is a highly complex process and models are in the early stages of refinement. Models have to describe the processes of advection through pressure gradients, entrainment, re-suspension and de-watering. Results vary enormously with different types of cohesive sediment.

4.4.4 Passive plume modelling

This is probably the most established of the models relating to plumes. The basic hydrodynamics can be well represented by a range of commercially available software. Of course, the models have to be site-specific if they are to provide relevant data for environmental impact assessment.

There are several types:

- random walk/advection/diffusion techniques
- finite difference approach
- finite element approach
- gaussian diffusion approach.

4.4.5 Water quality models

As with plume models, these tend to be add-ons to hydrodynamic models. Their purpose is to determine the fate of contaminants. They can also assess oxygen depletion due to the plume, but this is not frequently an issue because of the high degree of mixing that takes place. They usually include formulations for:

- adsorption/desorption
- precipitation/dissolution
- complexation/disassociation
- oxidation/reduction.

Contaminant behaviour is complex and highly site-specific. For any given situation there is considerable uncertainty in the description of the contaminant behaviour. The magnitude of this uncertainty is such that improvements in water quality prediction will only arise as the behaviour of contaminants under different environmental conditions becomes more closely defined.

5 Effects on the marine environment

5.1 INTRODUCTION

The environmental effects associated with sediment plumes are often secondary to those associated with the direct loss of seabed. While the loss of the seabed and its environmental resource is a fundamental and unavoidable consequence of dredging, the environmental effects associated with sediment plumes need not be.

This chapter reviews available information on the environmental effects of sediment plumes, focusing on water quality, marine ecology, fisheries and shellfisheries. It identifies and highlights those impacts potentially associated with sediment plumes arising from marine aggregate dredging and (separately) capital and maintenance dredging, indicating (where possible) the degree of significance associated with them. This is broadly summarised in Figure 5.1. It should be noted, however, that direct relationships between a single dredging operation and a particular impact are difficult to derive because of the site-specific nature of sediments and habitats and the variability of hydrodynamic conditions across time and space. Within reason, almost any impact could arise in any set of ircumstances, but most often potential impacts are predictable. Establishing "likelihood" and "significance" given the characteristics of the environment in question is important, therefore.

Potential cumulative effects caused by several dredging and non-dredging projects in the same area are considered separately. Environmental effects are related, as far as possible, to different dredging projects, sediment types, plume types and dredgers. Short-term and long-term effects are identified, highlighting how sediment plumes affect the water column and seabed in different ways and the influence of natural variability on this. Mitigation measures, including the concept of environmental windows, are also described.

5.1.1 Baseline environmental conditions

Environmental effects need to be assessed against the existing baseline environmental conditions at and around the dredging area. The baseline conditions must be relevant to the potential effects of the dredging project. An assessment of an individual or cumulative effect is only as credible as the baseline environmental data against which it has been judged. It is thus important to obtain or collect an accurate and representative dataset of the existing environment prior to the identification and prediction of environmental effects.

The credibility of a dataset is critical to predicting and monitoring the environmental effects of dredging. The baseline dataset must cover natural variations and seasonal patterns in order to provide the context within which to determine if a change constitutes an impact. The design of a baseline sampling programme is not a simplistic exercise. In particular, due consideration must be given to the term of sampling. One season or year of sampling may be inadequate to record the scope of inter-annual variation, eg if baseline surveys are limited to a period of exceptionally high or low benthic production, the true scale of impacts to benthos could be easily masked or misinterpreted.

```
                    ┌─────────────────┐
                    │ Sediment plumes │
                    └─────────────────┘
                             ⇩
```

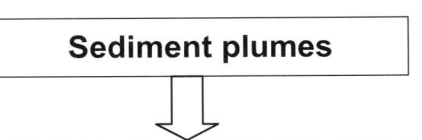

Effects on the marine environment

Aggregate dredging	Section	Capital and maintenance dredging
Reduced water transparency	5.2.2	*Reduced water transparency*
Contaminant mobilisation	5.2.3	*Contaminant mobilisation*
Increased oxygen demand	5.2.4	*Increased oxygen demand*
Effect on seabed communities	5.3.2 and 5.3.3	Effect on estuarine and shallow littoral benthic communities
Effects on juvenile and adult fish	5.4	Effects on juvenile and adult fish
Effects on fish migration	5.4.4	*Effects on fish migration*
Effects on shellfisheries	5.5	Effects on shellfisheries
Effects on sedimentology	5.6.1	*Effects on sedimentology*
Disruption of designated habitats/species	5.6.2	Disruption of designated habitats/species

Note: Italicised items indicate that, in general, only a limited risk exists

⇩

Cumulative effects
(Section 5.7)

⇩

Mitigation
(Section 6)

⇩

Monitoring

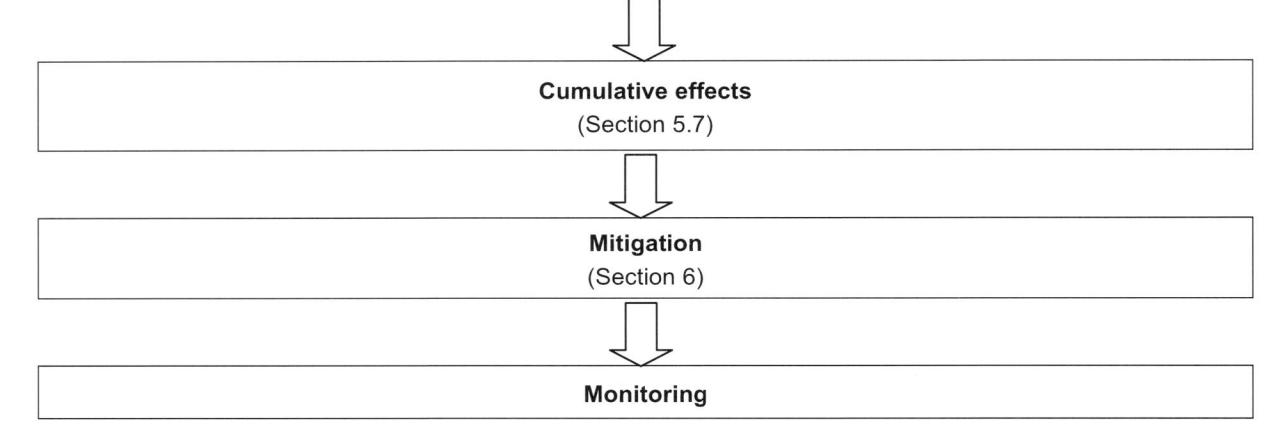

Figure 5.1 *A guide to effects on the marine environment*

In addition, natural variation in suspended sediment concentrations can occur on temporal scales that could confound comparisons with plume-related conditions during dredging if baseline data are collected inappropriately.

A variety of factors need to be investigated for the effects of dredging to be predicted, including a knowledge of existing water quality, biological communities, substratum, fisheries and shellfisheries resources. It is important to determine the temporal and spatial thresholds of acceptability in any particular environment (eg the tolerance of the species present). Tolerance thresholds need to be related to the environmental change caused by the re-suspension and movement of a sediment plume, particularly the concentration of re-suspended sediment or depth of sedimentation that can be tolerated over the background concentration. Such thresholds will be site-specific and species-specific. However, it might be possible to identify indicator species for different dredging environments.

As described in Section 5.4, some fish species are tolerant of turbid water conditions and so dredging-induced increases in turbidity might not cause a significant long-term effect. Therefore, baseline data need to reflect such tolerances to background conditions in order to determine the basis for assessing the significance of a plume-generated environmental effect.

In view of the above, the design of baseline sampling has to address the need for obtaining comparable data to assess impacts associated with a dredging-induced sediment plume. In the case of sediment plumes, which are temporary phenomena, it can be difficult to obtain baseline data that is suitable for assessing direct water column impacts in relation to natural variation. Accordingly, it can be difficult to obtain baseline data in the temporal sense and so spatial comparisons provide a more effective means for assessing and controlling impacts. Therefore, baseline data collection needs to take into account the approach to impact assessment. For example, it might be more effective to monitor benthic impacts arising from sediment plumes in terms of spatial impact rather than monitoring suspended sediment concentrations in the water column, the impact of which can be masked by a temporal component of natural variation.

As an example, baseline studies for the Øresund Fixed Link involved the Øresund-konsortiet carrying out extensive surveys of flora and fauna along the Link's alignment from 1992 to 1995. In addition to the surveys, investigations were undertaken to determine the individual plant and animal communities' vulnerability to construction work. The investigations focused on the effects of sediment spillage from dredging operations on benthic flora (eelgrass) and common mussels (Øresundkonsortiet, 1998).

In essence, to determine whether an effect constitutes an impact, information is needed first on the plume's concentration and its footprint, and second on the way this compares to background levels and the tolerance of the species present. All information should allow comparison between spatial and/or temporal changes to facilitate impact assessment and control. For example, tolerance thresholds alone cannot be used to determine impacts. In the case of dose-response relationships, tolerance has to be put into context with the spatial and temporal characteristics of the plume such as the concentration and duration of exposure.

5.1.2 Environmental changes caused by sediment plumes

The environmental effects associated with sediment plumes tend to occur as a result of two types of direct physical environmental change. Chemical changes can also occur if the sediment in the plume changes physicochemical conditions by reducing dissolved oxygen levels or introducing toxic contaminants to the marine environment.

The first physical change is associated with the presence of the sediment plume in the water column, which increases the concentration of suspended sediment in the affected water. The sediment type (ie cohesive or non-cohesive), dredger type (eg influence on plume position within the water column) and the dynamic or passive nature of the sediment plume influence the environmental effects associated with an increased concentration of suspended sediment. The source of the sediment is also important since it can determine the plume's physicochemical composition and contamination.

The second change occurs when the sediment plume settles out of suspension, thereby changing the environmental conditions of the seabed. Sedimentation can vary spatially in response to the nature of the sediment plume and bedload movement. Denser, deeper sedimentation might occur when a dynamic plume reaches the seabed compared to shallower, dispersed sedimentation from a passive plume.

Again, the sediment's physicochemical composition and contamination can be important. In the context of marine biological resources, the effect of sedimentation on the seabed depends greatly on the existing substratum and the ability of benthic life to either cope with or adapt to changed conditions.

5.1.3 Current knowledge about environmental effects

It should be noted that it is difficult to establish the environmental effects associated with sediment plumes without better field observations and research. For example, the effects of sediment plumes on gill abrasion in fish have been a persistent environmental concern in the United States, yet the actual effects remain unknown. Environmental studies pertaining to this phenonenom have been conducted in the laboratory, but using suspended sediment concentrations and exposure durations uncharacteristic of most dredging operations. This situation suggests that laboratory-generated data need to be verified against field observations. Generic gaps in existing knowledge should be borne in mind when reading the following sections.

5.2 EFFECTS ON WATER QUALITY

5.2.1 Introduction

Sediment plumes arising from dredging introduce sediment into the water column. Some waters contain naturally high ambient concentrations of suspended sediment and other particulate matter. Such concentrations can occur throughout the year or might occur seasonally as a result of soil run-off into estuaries during storms or increased algal biomass in coastal waters. Often the species in such waters are adapted to these conditions and can tolerate the short-term changes resulting from dredging-induced sediment plumes. However, concern increases where sediment plumes occur in waters that are normally relatively clear, where species require light penetration at depth or relatively sediment-free water in order to filter feed efficiently.

This section discusses the direct effects of re-suspended sediment on water quality. Three main water quality issues arise as a consequence of increasing suspended sediment concentrations in the water column: turbidity reduces light penetration, anoxic material utilises oxygen, and suspended concentrations can break water quality standards.

Changes in water quality can also have secondary environmental effects. Indirect effects on marine ecology, fisheries and shellfisheries, including effects associated with sedimentation, are discussed in subsequent sections.

5.2.2 Reduced water transparency

The re-suspension of sediment and other particulate matter into the overlying water column during dredging is often referred to as dredging-induced turbidity. Turbidity is not a measure of the suspended sediment concentration in the water column and it is important to recognise the differences. Turbidity is the interference with the passage of light rays through water caused by the presence of suspended matter scattering and absorbing light. Turbidity influences water transparency, limiting visibility in water.

Bathing water quality

European Union member states have to implement measures so that designated bathing waters comply with EC Directive 75/160/EEC. As well as microbiological parameters, water quality has to comply with standards for transparency. Compliance with this standard is waived for some bathing waters because of naturally turbid water conditions. For example, the bathing waters in the outer Severn Estuary, which separates south-west England and south Wales, are characterised by high levels of natural turbidity.

Water transparency effects during capital and maintenance dredging

Capital and maintenance dredging

Changes to background turbidity and transparency conditions due to sediment plumes are influenced by sediment type. Non-cohesive, fine sediment and organic matter will remain suspended in the water column for a longer duration than coarse sediment or fine cohesive sediment (eg compact clays) that quickly resettle.

The characteristics of the plume not only affect the duration of reduced water transparency, the physical characteristics of the sediment also directly affect turbidity and transparency, particularly because fine particles interfere with light more than coarse particles. For example, passage of light through several hundred parts per million of suspended clay and silt is significantly less than its passage through the same or higher concentrations of suspended sand.

This suggests that capital and maintenance dredging projects are more likely than aggregate dredging to affect water transparency. Capital and maintenance dredging activities generally tend to remove a greater proportion of finer material than aggregate dredging. In particular, maintenance dredging removes fine sediments settled in existing dredged areas. Unless dredging is undertaken using environmental dredgers that minimise re-suspension, the potential for sediment re-suspension is less if the fine material is cohesive or compacted, but greater if non-cohesive and easily mobilised.

An alternative situation arises with hydrodynamic dredging techniques, which are suitable for the removal of silts and, to a lesser extent, fine sands. The objective of some hydrodynamic techniques is to minimise the re-suspension of sediment throughout the entire water column since they rely on fluidising the sediment so it moves in a slurry just above the seabed. The effect on transparency throughout the rest of the water column can be minimal, although significant re-suspension higher up the water column can occur. This depends on the relative effects of injection rate, total water depth and the sediment's gas content. Where dredging intentionally re-suspends sediment throughout the water column, as for harrowing, reduced transparency is an inevitable consequence and its environmental effect needs to be considered before the dredge.

Water transparency effects during aggregate dredging

Aggregate dredging

Although aggregate dredging principally recovers coarse sediment, such as sand and gravel, the overflow contains mostly silt and clay. Generally, there is less potential for sediment plumes to affect water transparency significantly, except in cases where there is no screening of the overflow material. Coarse sediment resettles relatively quickly and the duration of reduced water transparency will normally be controlled by site-specific factors affecting sediment-settling velocities.

5.2.3 Mobilisation of contaminants

The mobilisation of contaminants is a problem most often associated with maintenance dredging, but occasionally arises with capital dredging. Contaminant mobilisation is not usually a problem associated with aggregate dredging. It is also fair to say that concern about contaminant mobilisation is more focused on disposal operations rather than dredging operations, particularly in terms of containment to prevent contaminants affecting the environment around the disposal site.

Pennekamp and Quaak (1990) reported that Dutch investigations into dredging-induced turbidity occurred in response to environmental concern about contaminants associated with fine sediment fractions in navigable waterways. Such concern led the Dutch Government to introduce legislation to reduce the contaminant loading from point sources such as effluent discharges. Nevertheless, local sediments still contained historic inputs of contaminants and it was recognised that dredging had the potential to release these contaminants into the water column, thus affecting the Dutch Government's policies to improve water quality.

Physicochemical environmental controls

Principally, the physicochemical environment controls the processes involved with the immobilisation and mobilisation of sediment-associated contaminants. The main sediment properties affecting the reaction of the sediment with contaminants are clay type and content, organic matter content, cation exchange capacity, reactive iron and manganese, oxidation-reduction potential (redox), pH and salinity. Of these properties, it is the clay, organic matter, pH change and redox conditions that predominantly influence the mobilisation of contaminants from the sediment.

Contaminant mobilisation occurs due to a dredging-induced change in physicochemical sediment conditions. Where dredging causes a sediment plume to arise, the physico-chemical environment changes considerably and substantial contaminant release can occur. This is not always the case; a change in the physicochemical environment can release contaminants from the sediment, yet favour other immobilising reactions.

Some contaminants pose a risk to the marine environment. The main contaminant groups are heavy metals, hydrocarbons and organochlorine compounds. Specific contaminants of environmental concern include tributyl tin (TBT), the biocide agent used in anti-fouling paint formulations. The environmental effect of each contaminant differs, depending upon the receiving environment, but contaminants are often discussed in terms of their toxicity, ability to bioaccumulate and environmental persistence. Tables 5.1 and 5.2 summarise the environmental effects associated with contaminants that are most often present in dredged material.

Heavy metals

Heavy metals are so termed because of their high atomic mass. Metals enter the aquatic environment from both natural and anthropogenic sources. Generally, trace amounts of metals arise from the weathering of rocks and soils. Natural contributions can be high in areas of metal ore bearing strata. Large quantities of metals enter the environment through diffuse sources such as run-off and atmospheric deposition in addition to point sources such as domestic and industrial wastewater discharges. Metals are used in many industries including manufacturing processes and as chemical catalysts.

Metals discharged into the naturally turbid estuarine water can be rapidly bound onto the surface of fine suspended sediment particles, by various adsorption processes. As the suspended sediment settles to the bed, the associated metals are gradually buried and become immobilised in anoxic sediment conditions. In many estuaries surrounding the North Sea, it has been estimated that about half of the metals entering an estuary become trapped within the estuarine sediments, with a lesser amount being eventually discharged to the sea (McLusky, 1989).

Metals (and other contaminants) are of concern because of their toxicity, persistence and tendency to bioaccumulate in living organisms. In addition to the amount of a metal present, toxicity depends upon the degree of its oxidation and the form(s) in which it occurs. The ionic form of a metal is generally the most toxic (eg cadmium 2^+). Toxicity can be increased if the metal is complexed with natural organic matter. Metallo-organic compounds such as methyl-mercury form under certain natural conditions and exhibit greater toxicity than inorganic elements alone.

A metal's ability to remain in the environment is known as its persistence. Unlike some organic substances (ie hydrocarbon and organochlorine compounds), metals tend not to decay at any appreciable rate and can remain indefinitely within the aquatic environment.

Aquatic organisms may bioaccumulate metals, depending upon the organism's physiology and the degree of metal bioavailability. Bioaccumulation is the ability of an organism to accumulate contaminants in body tissues. Depending on the degree of bioaccumulation and the sensitivity of the organism, accumulated contaminants can cause toxic effects such as tumours, bodily deformation and even death.

Table 5.1 summarises the environmental effects of a number of heavy metals often encountered in dredged material.

Table 5.1 *Environmental effects of heavy metals*

Contaminant	Environmental effects
Mercury	Toxic at low concentrations and more toxic if bacterially converted to methyl mercury; acute toxicity threshold of flounder is 1.5 μg/l (NRA, 1995); accumulates through the food chain affecting top predators.
Cadmium	Toxic at low concentrations but strongly adsorbed to sediment; acute toxicity threshold of Atlantic salmon is 156 μg/l (NRA, 1995); bioaccumulates in marine invertebrates and fish.
Lead	Varying toxicity depending on availability, uptake and species sensitivity (NRA, 1995) but can bioaccumulate in the food chain.
Arsenic	Moderate toxicity to fish and invertebrates and low bioaccumulation potential because of strong affinity to sediment, even in oxidised conditions.
Chromium, copper, nickel, zinc	Essential trace nutrients but potentially toxic at high concentrations; lethal concentration values for copper in sensitive aquatic organisms are usually <100 μg/l; zinc bioaccumulates in marine molluscs; nickel bioconcentration in marine organisms is generally low except for molluscs; chromium is moderately bioaccumulated and chromium III and VI differ in toxicity (NRA, 1995).

Hydrocarbon and organochlorine compounds

There are many types of hydrocarbon compounds and organochlorine (OCl) compounds that can adversely affect the marine environment. The following paragraphs describe the compounds most commonly occurring in dredged material, namely polynuclear aromatic hydrocarbons (PAHs), polychlorinated biphenyls (PCBs) and OCl pesticides.

PAHs are a group of organic substances containing two or more benzene rings. The types of PAHs of particular concern if present in the aquatic environment include naphthalene, fluoranthene and benzo(a)pyrene. Fossil fuels such as coal and oil are rich in PAHs, which are typically formed due to the incomplete combustion of organic material, including exhaust gases, or during mineral oil processing, aluminium smelting and coke production. In the aquatic environment, PAHs are found at low concentrations in water due to poor aqueous solubility. However, they are easily adsorbed to organic matter and inorganic particles in the water column and, should local sources exist, are likely to arise in deposited river silt.

PCBs comprise a skeleton of two linked benzene rings where one or more hydrogen sites are substituted by chlorine atoms. PCBs were manufactured for open-system uses (eg lubricants, paint additives) and closed systems (eg electrical transformers and electronic circuit condensers). There is no direct outdoor application for PCBs yet they enter the marine environment where they degrade very slowly. In the aquatic environment, PCBs tend to be adsorbed quickly by organic matter due to their hydrophobic nature.

Unlike many other contaminants, OCl pesticides are designed by manufacturers to be distributed in the environment, supposedly targeting a particular pest. Pesticides are often described in terms of their bio-concentration factor (BCF), which expresses the concentration found in organisms compared to the same substance in water. The following substances are commonly used in pesticide applications: aldrin, dieldrin, endrin, isodrin, DDT, hexachlorocyclohexane (HCH), pentachlorophenol (PCP) and hexachlorobenzene (HCB). Table 5.2 summarises the environmental effects of a number of hydrocarbon and organochlorine compounds.

Other contaminants

Of the various other contaminants that have the potential to affect the marine environment, tributyl tin (TBT) is often of concern during dredging projects. Since the discovery of its biocidal properties in the 1950s, the industrial application of TBT includes its use as the biocide agent in anti-fouling paints and coatings, molluscicides and agricultural fungicides. TBT enters estuaries from a limited number of point sources including dry docks and marinas, and many diffuse sources such as vessel hulls.

Because of its hydrophobic nature, once in the water column TBT readily comes out of solution and adsorbs to particulate matter and sediment. TBT also binds with phytoplankton, thereby introducing it to one of the lowest levels of the food chain.

Sub-lethal effects of TBT on non-target species were first observed in oysters during the 1970s (WWF, 1993). Continuing notoriety and investigation has shown that many species are sensitive to the toxic effects of TBT even at very low concentrations (Evans *et al*, 1995). TBT's effect on dog whelks and other gastropods is widely researched and reported. Most research has focused on an effect called imposex, where females develop male characteristics. It has been claimed that TBT was responsible for a decline of the dog whelk species *Nucella lapillus* in south-west England (Bryan *et al*, 1986).

Table 5.2 *Environmental effects of hydrocarbon and OCl compounds*

Contaminant	Environmental effects
PAHs	Bioavailability is often limited by PAHs affinity for sediment. Highest rates of bioaccumulation found in fish and shellfish, although only at concentrations lower than the surrounding sediment. Acute toxicity in aquatic organisms varies from 0.2 parts per million to 10 ppm (Donze (ed) 1990).
PCBs	Environmental persistence and lipophilic nature means that PCBs bioaccumulate in aquatic food chains such that bioconcentrations in top predators (ie seals) can be sufficiently toxic to initiate population decline. Acute toxicity in fish at concentrations exceeding 10 µg/l in marine water (NRA, 1995).
Aldrin	Developed as an insecticide but strongly adsorbs to solid particles in the aquatic environment because it is poorly soluble in water. Athough specifically toxic to insects, it can bioaccumulate in fish.
Dieldrin	A less toxic degradation product of aldrin and banned in Europe since 1981, but it is still used for combating termites in other countries; dangerous for fish since its persistence affords it time to bioaccumulate to toxic concentrations.
Endrin	Also banned in Europe since 1981 and resists biodegradation but is less toxic to fish.
DDT	DDT and its derivatives DDD and DDE are persistent and highly toxic to fish; despite this, it is still used in some countries to control malaria.
HCH (lindane)	The gamma isomer of HCH, known as lindane, is an effective insecticide often used as a general pesticide with application in agricultural and urban areas. Degradation is slow, facilitating bioaccumulation in aquatic organisms.
PCP	Extensively used as a wood preservative and is highly toxic in the marine environment; low aqueous solubility means that it adsorbs to sediments.
HCB	Used as a fungicide until 1975 when its agricultural application was banned in the UK. Very toxic to marine organisms and readily bioaccumulates.

Contaminant mobilisation between sediment and water

The risk of contaminant mobilisation affecting water quality and having subsequent environmental effects on aquatic life needs to be put into context with regard to the partitioning behaviour of individual contaminants. Contaminants have different degrees of solubility. Metals, such as lead, are quite insoluble and their partitioning from sediment is largely controlled by changes in pH. The potential for contaminant partitioning from the sediment to water can be measured through laboratory research. For example, sediment-water partition coefficients for TBT vary considerably, but are mainly are in the order of 10^3–10^4 (Waldock et al, 1990). It should, therefore, be recognised that many of the contaminants mobilised by dredging actually remain bound to re-suspended sediment rather than become dissolved into the surrounding water (limiting their potential impact). The environmental impact of mobilised contaminants is more of a concern after sediment plumes have settled on the seabed.

Contaminant mobilisation during capital and maintenance dredging

Capital and maintenance dredging

The physicochemical conditions favouring contaminant immobilisation occur in many estuaries and sheltered coastal areas that have historically received or currently receive significant inputs of contaminants emitted from land-based industry. The fine bed sediment in estuarine and sheltered coastal areas, including ports and access channels, is an effective "sink" for contaminants. Sediment in these areas can contain relatively high contents of clay and organic matter, to which contaminants become bound. Chemical and biological activity reduces the sediment's oxygen availability, creating conditions that favour contaminant immobilisation provided the sediment is not mixed, re-suspended and transported.

Contaminant mobilisation can occur due to sediment plumes generated during maintenance dredging and capital dredging. In particular, hydraulic dredging techniques employing trailing suction and cutter suction plant and some hydrodynamic dredging techniques such as harrowing have the potential to mobilise contaminants into the water column. The primary mobilisation effect of dredging is to change the redox state of the sediment by mixing it with large quantities of oxygenated water. Oxidation causes the destabilisation of contaminants bound to sediment. For example, as reduced sediment containing insoluble metal-sulphides is dispersed in oxygenated water during dredging, metals can be released when metal-sulphides are oxidised to sulphates.

PIANC Working Group 17 (1996) suggests that the amount of contaminated dredged material per project can be in an order of magnitude of 10–100 per cent for maintenance dredging (based on a 10 000 m^3 project) and 1–10 per cent for capital dredging (based on a 100 000 m^3 project). It therefore appears that maintenance dredging has more potential to mobilise contaminants, although capital dredging can mobilise a similar quantity of contaminants by virtue of larger dredging projects.

Contaminant mobilisation during aggregate dredging

Aggregate dredging

Contaminant mobilisation is much less of an environmental problem during aggregate dredging and capital dredging of coarse sediments. Coarse sediments such as sand and gravel contain relatively low amounts of clay and organic matter and are relatively less effective than fine sediment at immobilising contaminants. Aggregate sediments tend not to concentrate contaminants unless there is a nearby source of contamination, which is unlikely in open sea conditions.

5.2.4 Re-suspended sediment oxygen demand

Oxygen is essential to aquatic life. The oxygen content of natural waters varies with temperature, salinity and photosynthetic activity and respiration. Oxygen solubility decreases as temperature and salinity increase. Dissolved oxygen concentrations can vary seasonally and even daily, particularly in response to temperature and biological activity. Biological respiration, including decomposition of organic matter, reduces dissolved oxygen concentrations.

Oxygen demand of reduced re-suspended sediment

The surface sediment and interstitial water comprising the seabed are usually weakly to strongly oxidised and can contain measurable concentrations of dissolved oxygen. Abundant energy sources in the sediment (eg organic matter) support microbial activity, which subsequently causes oxygen deficiency as the available oxygen is depleted by respiration. When all the dissolved oxygen has been used the sediment conditions can become anoxic and microbes utilise oxygen present in nitrate and nitrite. Anaerobic conditions can arise and sulphate may be the only remaining source of oxygen. The continued utilisation of oxygen causes sediment, especially fine sediment, to become moderately to strongly reduced, even at depths of several centimetres below the surface. Coarser sediment, such as non-cohesive sand and gravel, is likely to be oxidised to deeper levels.

Dredging-induced sediment plumes can also affect dissolved oxygen in the water if the re-suspended material includes reduced anoxic and anaerobic sediment. The mixing of oxygenated water with reduced sediments can place a short-term but huge demand on the water's dissolved oxygen, potentially causing a rapid decline in concentrations. Brown and Clark (1968) have described secondary impacts of oxygen demand, for example on migratory fish (see Section 5.4.4).

Oxygen demand of organic matter and nutrients in re-suspended sediment

Sediment re-suspension due to dredging can place an unnatural demand on the dissolved oxygen by introducing organic matter and nutrients into the water column. Microbial activity (ie respiration) increases during the degradation of organic matter and in response to an increase in nutrients. This microbial activity is part of the water's self-purification process, but initially causes a rapid decline in the water's oxygen concentration. An unusually high oxygen demand causes an imbalance in natural cycles.

Oxygen demand during capital and maintenance dredging

Capital and maintenance dredging

In general, dredging-induced sediment plumes only pose a limited risk to water quality in terms of oxygen demand. Even if strongly reduced sediments are suspended, the affected water usually has the capacity to accommodate an increased oxygen demand; particularly where dredging occurs in the open sea or large estuaries. However, in some cases, there might only be a limited amount of water with which the re-suspended sediment can mix; for example, an enclosed harbour with limited tidal exchange. Concern of this nature arose during environmental investigations for capital dredging at Watchet Harbour in south-west England. Determinations of the sediment's *in situ* oxygen demand were made and predictions of the sediment's re-suspended oxygen demand were subsequently made (Posford Duvivier Environment, 1999). Using data based on S-factors (see Section 3.3.4) as a guide, it was estimated that the sediment re-suspended during dredging by cutter suction or backhoe techniques would not reduce dissolved oxygen to environmentally unacceptable concentrations.

Oxygen demand during aggregate dredging

Aggregate dredging

As sand and gravel is less likely to be strongly reduced and aggregate dredging typically takes place in the open sea, there is little potential for re-suspended sediment to pose a significant demand on dissolved oxygen levels. Reference to EIAs recently published in the UK suggests that oxygen demand is not considered to be an important environmental issue during aggregate dredging.

5.3 EFFECTS ON MARINE ECOLOGY

5.3.1 Introduction

The most significant impact on benthic marine ecology is the direct removal of sub-stratum and associated species, communities and habitats. Consequently, the majority of the literature relating to the impacts of dredging on marine ecology relates to direct physical removal of benthic fauna (see Van Moorsel and Waardenburg, 1990), although this literary situation might reflect the relative difficulty of researching dredging impacts on non-benthic species (eg highly mobile fish stocks). The effects of sediment plumes relating to increased suspended sediment concentrations and subsequent sedimentation are potentially the next most significant effects on marine ecology.

To review the potential effects of dredging-generated sediment plumes on marine ecology, it is necessary to identify the general nature of the benthic (bottom-dwelling) ecology that typically occurs in environments subject to dredging. Benthic communities present in different dredging areas have particular ecological characteristics (as detailed in Box 5.1). As a result, they respond differently to dredging-induced sediment plumes.

Benthic ecology affected by dredging

A number of potential impacts on marine ecology can arise as a result of the generation of sediment plumes associated with dredging. The ecological effects caused by aggregate dredging can be markedly different to equivalent effects caused by capital and maintenance dredging. The different responses of benthic communities to dredging arise because of the varying ecological nature of the benthic communities that occur in the dredging areas. The general coincidence of dredging areas with benthic ecology is summarised in Table 5.3.

Table 5.3 Coincidence of dredging with benthic communities

Benthic community	Dredging activity
Opportunistic (r-strategist) species present in dynamic estuarine and littoral habitats characterised by mobile fine sediment with salinity fluctuations	Capital and maintenance dredging of ports, harbours and their estuarine approaches
Competitive (K-strategist) species present in stable environments characterised by immobile sediment and consistent salinity	Offshore aggregate dredging and capital dredging of offshore approach channels
Intermediate communities recovering from natural or human disturbance characterised by dynamic and stable environments (r-strategist evolving to K-strategist)	Maintenance dredging (if reason for disturbance), possibly aggregate and capital dredging

Sections 5.3.2 and 5.3.3 describe the effects of sediment plumes on seabed and estuarine benthic communities respectively. Given the above, sediment plumes arising from aggregate dredging are more likely to affect seabed benthic communities and sediment plumes arising from capital and maintenance dredging are more likely to affect estuarine and littoral benthic communities. Sections 5.3.2 and 5.3.3 therefore focus on the effects of aggregate dredging and capital and maintenance dredging respectively.

Water column ecology

The following paragraphs provide a brief introduction to water column ecology, excluding free-swimming organisms (ie nekton) such as fish, cephalopods (eg squid and cuttlefish) and marine mammals. The consideration of phytoplankton is valuable since these organisms depend on light and so their distribution and abundance can be influenced by turbidity. Additionally, the presence of plankton can affect water transparency.

Box 5.1 *Benthic ecology – r and K strategists*

Opportunistic communities (r-strategists)

Biological communities show a gradient in terms of their composition that corresponds to an environmental gradient. Shallow sublittoral, littoral and estuarine environments are frequently subject to unpredictable short-term variations in environmental conditions, such as the disturbance that results from storm activity. The sediment composition in these environments is often highly mobile, comprising mud and fine sand, and the biological communities that inhabit such environments are adapted to these dynamic conditions (Newell et al, 1998). The populations are characterised by a high genetic variability that allows some components of the population to survive environmental extremes (Grassle and Grassle, 1974). They are known as opportunistic populations.

Opportunistic populations rely on a large investment in reproductive effort (rather than on mobility) for success in colonising new habitats made available by destruction of the previous community. They have particular life-cycle traits such as small size, high fecundity, rapid growth and high mortality. Opportunistic species are also known as r-strategists (Pianka, 1970).

Competitive communities (K-strategists)

Under more stable environmental conditions the community is controlled mainly by biological interactions rather than by extremes of environmental variability. The species that inhabit these environments are selected through maximum competitive ability in an environment that is already colonised by many species and in which space for settlement and subsequent growth is limiting. These species are called K-strategists and invest a larger proportion of their resources to non-reproductive processes such as growth, predator avoidance and investment in larger adults (McCall, 1976).

The recovery rate of biologically accommodated communities is dependent on the life cycle and growth rate of the particular community components. Some species, such as the ascidian *Dendrodoa* and the barnacle *Balanus*, may recolonise relatively rapidly, whereas larger slow-growing and long-lived bivalve molluscs, such as *Dosinia exoleta* or the mussel *Modiolus modiolus*, may take longer to recover following disturbance (Kenny and Rees, 1996, in Oakwood Environmental, 1997).

Intermediate communities (r- and K-strategists)

Finally, there are intermediate communities comprising species that combine weakened characteristics of the species occurring at both extremes of the environmental gradient. These communities have different relative proportions of opportunistic r-strategists and equilibrium K-strategists (Newell et al, 1998). Also the structure and physical size of the infauna (animals living within the sediment) changes along the gradient of environmental conditions. Changes have been described in relation to a number of impacts including organic pollution (Pearson and Rosenberg, 1978) and physical disturbance (Rhoads et al, 1978).

In terms of dredging, intermediate communities occur where species are re-colonising a dredged area or an area affected by dredging, such as an area smothered by the settlement of a plume. Opportunistic r-strategist species quickly colonise a disturbed area, although over time the benthos stabilises and competitive K-strategist species gradually colonise the same area. If regular disturbance occurs, which can occur with, for example, annual maintenance dredging, then this cycle will be repeated and an intermediate community will predominate.

In the marine environment, where the water depth is too great to allow penetration of light to the seabed, photosynthetic fixation by phytoplankton is responsible for the primary generation of organic compounds. Phytoplankton are tiny photosynthetic plants and usually consist of single-celled organisms or of chains of cells. In estuarine and coastal areas, the relative contribution of phytoplankton to total primary productivity is less than in the open seas because macroalgal communities, such as seaweeds, are able to survive as light can often penetrate to the seabed. Phytoplanktonic populations are less abundant in estuarine areas because the water column is often more turbid than in the sea. However, in the sea, light becomes a limiting factor for photosynthesis at a particular depth, so other factors combine to control the rate of phytoplanktonic photosythesis. These factors include turbulence, nutrient concentration and grazing.

Zooplankton are non-photosynthetic planktonic animals and range from single-celled forms to smaller vertebrates such as larval fish. Zooplankton graze on phytoplankton and therefore zooplankton distribution tends to mirror that of the phytoplankton. Since phytoplankton congregate in surface waters because of their requirement for light, this is also where the majority of the zooplankton are found.

5.3.2 Effects on seabed benthic communities

Aggregate dredging

The impacts of the deposition of sediment from plumes generated due to the dredging of sublittoral sandbanks in Moreton Bay, Australia, has been studied by Poiner and Kennedy (1984). They showed that the population density and species composition of benthic invertebrates adjacent to a dredging area (ie potentially the zone of influence of the plume) was much greater than that within the dredged area. This conclusion may be expected considering the small amounts of sediment that are generally deposited outside of the immediate vicinity of the dredged area in the case of aggregate dredging.

Other studies on the sedimentation of particulate material also suggest that sedimentation is mainly confined to a zone of a few hundred metres from the discharge chute (Newell *et al*, 1998).

Should significant deposition of sand occur in areas that have a similar sediment type, the impacts on benthic communities are likely to be small. The benthic community in such an environment is likely to be adapted morphologically and behaviourally to a dynamic environment. It is, therefore, likely to be able to cope with the disturbance caused by sedimentation.

Generally, there is little evidence to suggest that the deposition of sediment outside dredging areas can have a significant effect on benthic communities (Newell *et al*, 1998). However, this results from the lack of targeted field evidence to demonstrate whether this is the case, rather than certainty regarding the level and nature of the impact. One explanation for the potentially small effect on marine ecology from dredging-induced sediment plumes is that many benthic species are able to tolerate high levels of sediment deposition. Also several major species groups, for example polychaetes, molluscs and crustacea, can successfully recolonise areas which have been subject to sedimentation by undergoing vertical migration (Maurer *et al*, 1986). The effects of elevated turbidity and sedimentation are likely to be more significant in marine environments that have naturally low background concentrations of suspended sediment and deposition rates. Benthic communities in these environments are likely to be more sensitive to conditions of elevated suspended sediment and turbidity.

Despite the general view that the scale of any aggregate dredging-induced sediment plume effects on marine ecology is limited, the potential for adverse effects is high under certain circumstances, particularly in environments dominated by gravel substrata. In gravel seabed environments, sedimentation is most likely to affect sessile species because they are unable to burrow or vertically migrate in response to an increased sedimentation rate. Sessile species include delicate organisms such as bryozoans and hydrozoans (Emu Environmental, 1998).

Sedimentation also affects species whose functioning is inhibited by suspended sediment such as filter-feeding epifauna, for example sponges. Filter feeders are particularly susceptible to a significant increase in suspended sediment because their feeding apparatus could become overloaded. Coral and kelp forest communities are also susceptible to increased sedimentation rates (Hodgson, 1994 in Selby and Ooms, 1996).

Benthic communities in more stable, relatively undisturbed environments are biologically controlled and are dominated by populations of species adapted with K-strategist ecological characteristics. These communities form in areas of hard substrata, such as gravel and rock, which provide the conditions for a stable biological community. Again, the generation of a significant sediment plume and an increase in sedimentation could affect the benthic community. For example, turbidity and sedimentation can affect short faunal turf communities, with the hydroid genera *Turbularia* and *Obelia* being more tolerant than other hydrozoan species to turbid waters (Oakwood Environmental, 1997).

Box 5.2   *Effects of sediment plume settlement (ERM, 1999)*

A prospecting licence was issued to Westminster Gravels Ltd in 1997 to identify sand resources at the Middle Bank, Firth of Forth. The subsequent production licence application required the preparation of an environmental statement that assessed the potential effects of the extraction of sand on marine ecology.

This example indicates that benthic communities are susceptible to dredging-induced sediment plumes even during aggregate dredging should fine deposits overlie sand and gravel. This was the situation in the Firth of Forth, where the aggregate resource was covered with an overburden which comprised, on average, 19 per cent shell gravel, 22 per cent silt and 59 per cent sand. Whilst a sediment plume may migrate outside the immediate dredge area, the effect on benthic communities depends on their sensitivity to disturbance.

The environmental statement concluded that the major impact on marine ecology would be the direct impact at the site of the aggregate dredging. However, the deposition of sediment in the vicinity of the dredge site would cause the smothering of benthic species, with the degree of impact depending on the degree of sedimentation and the sensitivity of the species. In particular, sessile species, such as tube dwellers and filter-feeding epifauna, would be most susceptible and would be expected to be lost from areas with significant sediment deposition. Some species of feeding bivalves are nevertheless highly adaptable to feeding in turbid environments.

The loss of key species in these communities, such as *Sabellaria* spp, can lead to the collapse of the entire biologically accommodated community even though individual species within the community may be apparently tolerant of environmental disturbance (Newell *et al,* 1998). The polychaete worms *Sabellaria spinulosa* and *Sabellaria alveolata* are colonial and can form extensive reef structures on rock, gravel and cobble substrata. These reefs modify the habitat and create an environment that is more complex, with a greater diversity of micro-habitats, than was previously present. Diverse communities of epifauna are often associated with *Sabellaria* spp reefs as a result. *Sabellaria* spp reefs undergo natural cycles of accretion and decay along with the associated community with a periodicity of five to ten years (Gruet, 1986). Should excessive sedimentation adversely affect a community, the recovery of the community to its pre-disturbance condition may take several years, although the *Sabellaria* spp reefs may recover relatively rapidly.

The effect of a sediment plume is likely to be less significant during dredging at locations where the levels of suspended sediments are naturally high. In this type of environment the benthic community is adapted to the natural turbidity. For example, the extraction of sand from the Helwick Bank in the Bristol Channel would potentially generate a limited sediment plume due to the low proportion of fines in the sediment. Yet it was concluded that the potential impact of the plume would be minimal due to the high levels of suspended sediment that occur naturally in the Bristol Channel (Gibb, 1997). Nevertheless, such conclusions need to be substantiated through the determination of thresholds to turbidity and high suspended sediment loads and supported by targeted field investigation. The example described above suggests that thresholds need to be set with regard to background concentrations of suspended sediment, particularly high natural concentrations.

Increased turbidity resulting from dredging activities may depress phytoplankton productivity by reducing light penetration into the water column. This effect is localised around the dredger and insignificant in terms of total production (de Groot, 1986).

5.3.3 Effects on estuarine and shallow littoral benthic communities

Capital and maintenance dredging

Capital and maintenance dredging operations occur largely in estuaries and harbours, often adjacent to shallow littoral areas. Estuaries are usually subject to frequent deposition and re-suspension of fines and are therefore naturally turbid environments (SOAEFD, 1996). Many of the macrofauna that live in areas of sediment disturbance are well adapted to burrow back to the surface following burial (Schafer, 1972). For example, some benthic animals are able to migrate vertically through more than 30 cm of deposited sediment, and this ability may be widespread even in relatively deep waters (Maurer *et al*, 1979). Such communities are therefore unlikely to be significantly impacted by increased sedimentation due to dredging-derived material. This will also apply for areas that are subject to natural disturbance from, for example, storms.

The ability of many estuarine communities to cope with disturbance, such as excessive sedimentation from a sediment plume, can be explained by their opportunistic (r-strategist) life-history strategy, as described in Section 5.3.1.

Despite the ability of some individual species within benthic estuarine communities to cope with relatively high levels of sedimentation, other species and communities are likely to be affected by sedimentation, at least in the short-term. The overall area affected by sediment plumes in estuarine areas can also be relatively large.

For example, in eastern Hillsborough Bay, Florida, a silt plume covering three square miles has been detected during clamshell (ie grab) dredging operations (Simon and Dyer, 1972). Immediate effects on benthic communities that can occur include the removal of organisms by suffocation, with gross reductions in faunal biomass, number of species, and abundance (for example, Kaplan *et al*, 1975). However, as discussed previously, recolonisation after such a disturbance is likely to be rapid due to the opportunistic nature of many estuarine benthic species.

A change in the nature of the sediment composition, as would result from the deposition of silt onto an area of coarser sediment, may have an impact on benthic community structure. The larvae of many species of benthic invertebrates are sensitive to sediment composition and will not metamorphose from the planktonic stage until contact is made with a suitable substratum (Taylor and Saloman, 1968).

Sedimentation of material can have implications for the sediment chemistry, which in turn can have effects on benthic communities. Fine sediments can form a thin surface ooze, which has a restricted oxygen circulation within it and can lead to oxygen reduction within the surface sediments and the overlying water column (USEPA, 1976).

In addition to the impacts on typical estuarine benthic communities described above, the generation of sediment plumes due to capital and maintenance dredging within estuaries has the potential to affect the more sensitive species that may be present in estuaries. As the species are more sensitive, impacts may be judged as being more significant. One such community is the maerl community.

Maerl beds form in estuarine areas where conditions are relatively sheltered but where there is still sufficient water movement to keep the community free of significant depositions of fine particles. Maerl beds are largely formed from two species of coralline algae, namely *Lithothamnium corallioides* and *Phymatolithon calcareum*. These species are listed in Annex V of the Habitats Directive.

Additionally, there is often a diverse epifaunal community associated with maerl. For example, extensive maerl beds are found within the Fal and Helford candidate Special Area of Conservation (cSAC) in the UK.

The generation and subsequent deposition of significantly higher levels of suspended sediment than background levels would have a detrimental effect on the maerl community. The smothering effect and the consequent reduction in light penetration could kill off the maerl. This effect may be long term, even if the dredging activity were to cease. This is because the algal species making up the maerl community are slow-growing, with growth rates of 1–2 mm per year being observed (Adey and McKibbin, 1970), and therefore full recovery would take a long time.

5.3.4 Effects on water column communities

Reduced photosynthesis

Capital and maintenance dredging

Many estuaries exhibit naturally occurring turbid water conditions. However, capital and maintenance dredging can increase these levels beyond environmentally tolerable limits. By decreasing the penetration of light into the water column, primary productivity can be diminished and the rate of respiration of phytoplankton can exceed the rate of photosynthesis (Johnston, 1981).

The rate of photosynthesis by primary producers (such as macroalgae and phytoplankton) in marine waters is affected by a number of factors, one of which is the rate of supply of light energy in terms of either irradiance or intensity (Barnes and Hughes, 1988). As light intensity increases from zero, there is an increase in the rate of photosynthesis up to a plateau phase, after which photosynthetic rate decreases at higher light intensities.

In the surface waters of the sea, photosynthetic production varies with light intensity and also with depth because light intensity decreases exponentially with depth. At the water surface, light intensity may be high, thus causing photo inhibition (a decreased rate of photosynthesis). Below this depth, a point is reached where light is not limiting photosynthesis but where photo-inhibition does not occur. This is the region of maximum photosynthetic rate. Below this depth, light becomes a limiting factor and the rate of photosynthesis declines.

In contrast to this, respiration rate does not vary with depth. Therefore, a depth is reached where photosynthetic rate and respiration rate are equal (the compensation point). Above the compensation point, primary producers (eg phytoplankton) can grow and multiply. Below this depth they must subsist on accumulated reserves.

An increase in the turbidity of the water column due to dredging-induced sediment re-suspension can reduce water transparency, thereby decreasing the penetration of light through the water column (see Section 5.2.2). This effect can cause the compensation depth to be shallower than in less turbid waters and reduce the total primary production in the water column.

Turbidity increases can also affect macroalgae attached to the seabed by reducing the compensation depth. A decline in the algae population can occur if the compensation depth were to be re-positioned above the region of algae growth because the rate of respiration would be greater than the rate of photosynthesis. Under extreme conditions, this effect can potentially eliminate algae dominated communities.

Excessive turbidity can also result in the destruction of attached algae and vegetation rooted to the seabed, thus eliminating the food base of estuarine ecosystems (USEPA, 1976). It has been shown that reduced light penetration in shallow estuaries can cause a reduction in seagrass populations by inhibiting their ability to photosynthesise effectively (Simon and Dyer, 1972). However, care needs to be taken when addressing the effects of reduced photosynthesis. In the USA, the dredging-induced effects on seagrasses remains controversial (eg Onuf, 1994 and Long *et al*, 1996).

5.4 EFFECTS ON FISH

5.4.1 Introduction

Fish are potentially sensitive to sediment plumes in terms of suspended sediment in the water column and sediment settlement on the seabed. The severity of the sediment plume effect on a particular species depends largely on the sensitivity of the species and the background conditions that prevail in the environments in which dredging takes place. Since aggregate dredging usually occurs in offshore waters that are often subject to naturally lower turbidity levels than harbours and estuaries, the potential for fish species to be affected by sediment plumes is greater in the case of aggregate dredging than for capital and maintenance dredging.

There are, however, exceptions to this simplistic view; one such example being the Bristol Channel, where aggregate extraction occurs in naturally turbid waters. Capital and maintenance dredging have most potential to damage sport and commercial fisheries, where fish species live in estuaries during part or all of their life cycles, including migration (Taylor and Saloman, 1968).

The following paragraphs describe the effects of sediment plumes on fish. The effects can be divided into effects due to a higher suspended load in the water column and effects due to the settlement of the suspended load on to the seabed. The effects of sediment plumes are described in relation to their impacts on the biology of individual species. Specific examples are used to illustrate such effects.

5.4.2 Effects on fish development

Effects on fish larvae

Sediment plumes may affect fish during their early life stages. For example, pelagic fish larvae and eggs concentrate in the ocean's surface layers and are susceptible to the effects of dredging-induced sediment plumes, including the cold water overflowing from the dredger.

Generally, the effects of dredging-induced sediment plumes on fish larvae have received little study within EIAs for aggregate dredging and capital and maintenance dredging projects in the UK. Similarly, this subject appears to have been given little consideration overseas. However, there is evidence from studies of other industry-induced sediment plumes that suggests that fish larvae avoid turbid waters. For example, Miller (1974, in MMS, 1993) studied fish larvae densities in waters off sugar mills, sewage outfalls, oil refineries, thermal outfalls, urban discharges and harbours. It was reported that turbid waters reduced larvae species and larvae density and abundance by 75 per cent and 55 per cent respectively. Overall, turbidity, whether natural or man-made, was negatively correlated with larval fish abundance.

In the United States, research has been carried out into the environmental effects of marine mining (for example, MMS, 1993). Marine mining is defined here as the extraction of aggregates and other minerals from the seabed. Marine mining creates sediment plumes potentially similar to those induced by aggregate dredging and offshore capital dredging. Although it is difficult to establish how applicable the effects of mining-induced sediment plumes are to dredging projects, research in this field provides information about the generic effects of sediment plumes.

Matsumoto (1984, in MMS, 1993) studied the effects of sediment plumes arising from marine mining on tuna and billfish larvae, indicating that the normal development of pelagic fish embryos, larvae and post-larvae in the ocean's surface layers could be affected by marine mining. In particular, fish larvae could be affected by sedimentation, loss of illumination due to turbidity and changes in the property of water, such as temperature. Reduced illumination caused by turbidity was reported to affect feeding by reducing the reactive distance and feeding efficiency of fish larvae. Some species have overcome the effects of sediment plumes by increasing their swimming speed and searching area for prey. Matsumoto reported that tuna and billfish larvae could compensate for reduced illumination better than most fish larvae because of their exceptionally large eyes and mouth and rapid fin development.

The direct effects of suspended solids on the early developmental stages of fish include blocked food intake and clogged gills. Yet Matsumoto suggested that such effects generally occur at much higher turbidities and after much longer exposures than anticipated for the marine mining induced sediment plumes. Accordingly, limited exposure should not cause ill effects during the critical early life stages (Matsumoto, 1984, in MMS, 1993).

Effects on fish spawning

Fish spawning is affected where the settlement of sediment plumes directly smothers fish eggs or changes the composition of the substratum of a nursery area. In the USA, the EPA has speculated that siltation within estuaries adversely effects fisheries by eliminating food supplies, destroying habitat and eliminating spawning areas or smothering eggs and larvae after spawning has been completed (USEPA, 1976). Destruction of demersal eggs by siltation is of particular concern in upper estuarine nursery areas (Sherk, 1971), a potential consequence of capital and maintenance dredging.

Experiments by Westerberg *et al* (1996) tested the effects of the exposure of cod eggs to sediment suspensions. It was concluded that adhering particles gave a loss of buoyancy that was proportional to the sediment concentration and the exposure time. Furthermore, the process is purely mechanical and will apply to all pelagic eggs, regardless of species. If an egg sinks to the seabed, this will in most cases mean a certain loss to benthic predation, so the process will have a direct effect on reproduction success (Westerberg *et al,* 1996).

Fish species requiring a particular sediment type on which to spawn are perhaps most susceptible to changes in sediment composition that may arise due to the settlement of material suspended in sediment plumes. Certain species, for example the herring *Clupea harengus*, lay demersal eggs that adhere to stones or gravel (or algae). A change in the structure of the spawning ground, such as an increase in the fines content on the substratum, may prevent the eggs from adhering to the coarse bed material. Given that herring spawning beds are small and that they select specific gravel beds year after year even within a large gravel area, settlement of fines from sediment plumes can significantly affect their successful reproduction (ICES, 1992).

Sandeels are a major target of the UK fishing industry and are susceptible to the effects of dredging. Sandeels lay their eggs in sand and sand grains of a certain size adhere to them. When the eggs are fully covered with fine sediment, the development of the embryo will be arrested. If the outwash fines released during dredging settle on sandeel eggs, the smothering by fine sediment may cause less successful hatching (ICES, 1992). In addition, some species such as herring, dogfish, skates/rays, lobster and edible crab release their eggs at the seabed. The eggs of these species are also vulnerable to damage due to the settlement of sediment plumes (Oakwood Environmental, 1999).

Water from the area adjacent to the seabed may be colder than surface water although this temperature depends on local circulation patterns and water depth. However, fish eggs and larvae can be adversely affected if they came into direct contact with the colder discharge water (MMS, 1993). Any effect is likely to be localised because discharge water and surface water mix rapidly. The significance of any effect depends on site-specific factors such as the temperature differential between the surface and discharge waters and the local mixing regime. For example, if the bottom waters were much colder than the surface waters, the denser cold water would only remain at the surface for a short period of time. Matsumoto (1984, in MMS, 1993) reports that the contact of surface-drifting larvae and eggs with colder water can:

- cause the cessation of embryonic development
- cause the development of deformed larvae
- result in thermal shock to the larvae, causing them to lose equilibrium or become easy targets to predators
- result in death.

Overall, there appears to be a lack of field evidence to support the assessment of sediment plumes on the survival of fish spawn, particularly the potential differences between pelagic fish spawning over extensive sea areas compared with demersal fish spawning at selected sites. The assessment of dredging-induced sediment plumes needs to be assessed against field evidence of the different spawning areas within and around a dredging area and the effects of suspended sediment and/or sedimentation on hatching rates.

Effects on fish development during aggregate dredging and capital and maintenance dredging

Capital, maintenance and aggregate dredging

Many of the generic effects on juvenile fisheries associated with sediment plumes are similar for both aggregate dredging and capital and maintenance dredging. Therefore a detailed description of the different effects on fish development due to sediment plumes arising from the different types of dredging is not necessary.

5.4.3 Effects on adult fish

Effects on fish health

Coarse particles in suspension may harm fish by abrading the body surface and possibly removing protective mucus, thereby increasing susceptibility to invasion by parasites or disease (Everhart and Duchrow, 1970). Very high levels of suspended sediment can suffocate fish by clogging their gills and filling the opercular cavity. However, it is likely that fish would avoid such conditions prior to these adverse impacts on health.

In the USA, EPA studies near an active dredging site recorded a variety of effects on fish including a reduction in general swimming activity. Fish were observed also to engage in "coughing" and gill-scraping behaviour in an attempt to free the gills of accumulated particulate material (USEPA, 1976). Other studies have hypothesised that the re-suspension of bed material by dredging operations during warm months may be lethal, especially during periods of maximum fish migration (Brown and Clark, 1968, see Section 5.4.4).

Research by Westerberg et al (1996) suggests that the primary response of free-swimming adult herring and cod is to avoid areas of elevated suspended sediments. This study was carried out as part of the environmental impact assessment for the Øresund Link project (see Box 5.3).

Effects on fish feeding

In particular, visual feeders such as tuna, mackerel and turbot might avoid turbid water if it would impair their ability to locate prey. The following paragraphs describe how sediment plumes might affect fish that feed using vision to locate prey.

The effects of sediment plumes on visually feeding fish species have been little studied in the UK, although overseas examples exist. Records demonstrate that tuna are usually caught in clear waters and rarely in turbid waters, suggesting that they perceive their prey better in clear waters (Bane, 1961 in MMS, 1993).

Box 5.3 *Effects of suspended sediments on the behaviour of adult herring and cod*

> Westerberg et al (1996) carried out experiments on the avoidance behaviour of adult cod and herring. These experiments were made in a large saltwater flume, divided for part of its length by a separating wall. Suspended sediment was injected to one half of the flume and the concentration of suspended sediment in the flume was varied between experiments. Experiments were conducted using both clay and lime. The fish were observed and their positions were recorded every 15 seconds for a period of 20 minutes.
>
> From these experiments, it was concluded that the adult fish displayed avoidance behaviour at a very distinct threshold at an unexpectedly low sediment concentration. This threshold was essentially the same for both the sediment types tested. Expressed as turbidity, the threshold was approximately 5 NTU and 3 mg/l in terms of sediment concentration.
>
> In all of the experiments it was noted that the fish had an initial awareness reaction and withdrew from the channel where the sediment addition was just beginning. This reaction usually occurred well before any sediment clouding could be seen, at a much lower concentration than in the final plume.

Barry's (1978, in MMS, 1993) study of captive yellowfish tuna in turbid waters observed no ill effects from short-term exposure to suspended solids concentrations ranging from 9 mg/l to 59 mg/l. However, the tuna were sometimes observed to avoid the turbid cloud and stop feeding in turbid waters, but they also readily passed through the cloud (Barry, 1978 in MMS, 1993). Barry advised that some caution should be taken when extending these observations to tuna in the wild. While the captive tuna may have developed higher tolerances to pre-experimental turbidity in the tank, the ability of tuna to detect turbid water in the wild might differ from that of the captive tuna.

According to the above, sediment plumes arising from dredging might change the behaviour of fish that target their prey visually. Fish that normally inhabit relatively clear water are more likely to have their feeding patterns impaired by dredging-induced sediment plumes than fish that inhabit naturally turbid waters.

As described in Section 5.3, benthic communities may be smothered as sediment plumes settle on the seabed. This effect might cause the burial and/or death of the food resource for fish species that feed on benthic organisms. Alternatively, the physical composition of the substratum may be altered, changing the benthic ecosystem. This may potentially affect fish species if a major prey item is removed from the invertebrate community. Overall, in particular cases some fish may become starved of their primary food source, although it is more likely that the fish growth rate would be reduced or the fish would move away from the affected area (Oakwood Environmental, 1999).

Effects on adult fish during aggregate dredging and capital and maintenance dredging

Capital, maintenance and aggregate dredging

As for juvenile fish, many of the generic effects on adult fisheries associated with sediment plumes are similar for both aggregate dredging and capital and maintenance dredging. Therefore a detailed description of the different effects arising is not necessary.

5.4.4 Effects on fish migration

The potential for sediment plumes to affect fish migration depends on the duration of the dredging activity and timing. A short-term or intermittent plume may not disrupt migration if sufficient time is available between plumes. Migration is seasonal so an impact would be more significant if the dredging were more intensive and occurred at a critical period during the migration.

Under Annex II of the EC Habitats Directive, there are several protected fish species that can be affected by sediment plumes during their migration. These species include sea lamprey, allis shad, twaite shad and river lamprey. Population declines of these species are attributed to various causes including obstructions to migratory routes.

Other studies have concluded that the re-suspension of bottom material by dredging operations during warm months may be lethal, especially during periods of maximum fish migration (Brown and Clark, 1968). This impact can arise as chemical and biological activity is greater during warmer periods, which increase the oxygen demand on the water column. The potential increase in oxygen demand, which can arise due to the re-suspension of bottom sediments, may exacerbate this situation. During migratory periods, the oxygen requirement of migratory fish is likely to increase due to their greater activity. Therefore low oxygen levels during these periods are perhaps more likely to result in adverse impacts on fish than at other less critical times. Since the intermittent nature of dredging operations is unlikely to result in prolonged periods of low oxygen conditions, the potential for significant adverse effects on migratory fish is low.

Migration disruption during capital and maintenance dredging

There is little information available relating to the effect of capital and maintenance dredging on fish migration. However, capital and maintenance dredging can occur in important migratory routes such as estuary mouths and along the lower reaches of rivers. In particular, sediment plumes can inhibit the ability of fish to locate the mouth of estuaries on their migrations from the sea to rivers and/or block the migratory route entirely. Studies for the Øresund Link project noted that high concentrations of suspended sediment can block the migration of certain fish and might affect species that migrated through this area between March and October, including eel, garfish and lumpfish (Øresundkonsortiet, 1997).

The potential for sediment plumes to affect fish migration has been the subject of various studies. Despite the detail of the research carried out by Palermo *et al* (1990), Environment Canada (1994) concludes that such research cannot determine that dredging does not have a major impact on fish migration. It is therefore suggested that the impact of sediment plumes on fish migration requires further research, including targeted fieldwork.

Migration disruption during aggregate dredging

Aggregate dredging

The potential for aggregate dredging to affect fish migration appears to be limited since most dredging takes place in open water where there is sufficient space for fish to pass around a sediment plume. However, the impact of sediment plumes from sand extraction from the Middle Bank, Firth of Forth, on migratory fish species was considered by ERM (1999). It was considered that there was a potential impact in the disturbance to migrating salmon and sea trout due to increased suspended sediment levels and decreased oxygen levels in the water column (ERM, 1999). This potential impact is more likely to arise in this situation as the aggregate dredging was taking place in a relatively enclosed body of water in comparison to the open sea. In addition, dredging that affects a long-term passive plume is more likely to affect the entire water column.

It was concluded that fish tend to avoid areas of high suspended solids and low oxygen concentrations. The peak level of suspended sediment concentration in the vicinity of the licence area was predicted to reach a level that might cause mild physiological stress in salmonids. However, this was expected to be short-lived and the mean level would be lower. Also, significant decreases in dissolved oxygen would be confined to the centre of the plume. As the plume would not form a barrier across the Forth, it would not affect salmonid migration.

5.5 EFFECTS ON SHELLFISH

5.5.1 Introduction

Studies have identified the potential for shellfish grounds being damaged by sedimentation resulting from aggregate dredging. For example, the extraction of sand and gravel from Nash Bank in the Bristol Channel was reported to have the potential to affect shellfish grounds due to smothering with sediment (BMT, 1996).

In November 1990, Civil and Marine made a formal application for a production licence to extract sand and gravel from the Bristol Channel Outer Area 394. The environmental statement for this application proposed the possibility of the loss of lobster habitat and potting grounds due to the deposition of suspended sediment, both inside the licence area and in an area along the tidal axis (Oakwood Environmental, 1993).

Similarly, some studies on the effects of sedimentation on shellfish have concentrated on the deposition of fine sediments, which are more characteristic of estuarine areas. Therefore, these effects are more likely to occur during capital and maintenance dredging operations.

5.5.2 Effects on shellfish habitat

The settlement of a sediment plume can affect epifaunal crustaceans. Lobsters and crabs can be at risk from habitat loss due to the silting up of seabed crevices in which they live. A localised redistribution of crustaceans can occur as a reaction to habitat loss. However, this effect, along with the potential for the silting up of pots, can lead to declining catches from traditional fishing grounds. Edible crabs, which become torpid while brooding, are especially susceptible to smothering and suffocation by sediment fallout (Howard, 1982 in ICES, 1992).

Indirect effects could arise should benthic prey habitats be smothered or benthic populations are modified as a result of a change in the sediment composition. However, some studies have concluded that the limited near-field dispersion and short-term duration of significant sediment plumes from aggregate dredging only causes localised direct impacts on shellfish.

Effects on shellfish habitat during aggregate dredging and capital and maintenance dredging

Capital, maintenance and aggregate dredging

As for fisheries, many of the generic effects on shellfish habitats associated with sediment plumes are similar for both aggregate dredging and capital and maintenance dredging. A detailed description of the different effects arising is thus not necessary.

5.5.3 Effects on shellfish species

Filter feeding

Shellfish, eg the mussel *Mytilus edulis*, are filter-feeding bivalves (but note that in the USA crab and shrimp, non-filter-feeding species, are also considered to be shellfish). These species help to control natural phytoplankton and seston (particulate matter suspended in seawater) loads in the water column to an extent that food for the bivalves may become a limiting resource in the benthic boundary layer at the sediment-water interface (Snelgrove and Butman, 1994).

This high efficiency at removing particulate matter from the water column creates direct adverse effects of dredging-induced sediment plumes on filter-feeding bivalves. Large quantities of suspended sediment in the water column can cause the loss of filter-feeding components of the community through clogging of the gills (Newell *et al*, 1998). Furthermore, suspended sediment can also be detrimental to filter-feeding organisms due to impairment of proper respiratory and excretory functioning and feeding activity (Sherk, 1971).

Overall, research suggests that the degree of an impact on filter-feeding species depends on the extent of sedimentation against background levels. Benthic-dwelling shellfish in normally turbid environments are probably able to survive smothering unless the rate of deposition is excessive, whereas less tolerant species may be excluded (ICES, 1975). Studies also show that filter feeders, and bivalves in particular, are highly adaptable in their response to increased suspended sediment levels from, for example, periodic storms and dredging. Filter feeders maintain their feeding activity over a wide range of phytoplankton concentrations and inorganic particulate loads (Newell *et al*, 1989).

Shellfish health

The dredging of certain sediments, particularly in estuarine areas, has the potential to cause the development of toxic algal blooms, which may subsequently affect shellfish health. Toxic algal blooms resulting in paralytic shellfish poisoning have been recorded in some locations, for example in the Firth of Forth, and are mainly attributed to the dinoflagellate *Alexandrium tamarense*. The sediments in such areas host the resting stages, or cysts, of this species of plankton. When cysts occur in large numbers they have the potential to seed toxic blooms, which damage the shellfish industry, pose a risk to human health and may affect seabirds (Ayres and Cullum, 1978 in ERM, 1999). Disturbance of such sediments through dredging could lead to the development and spread of toxic algal blooms.

Studies have shown that high levels of suspended sediment can decrease oyster growth by reducing pumping rates. Pumping provides for respiration and nutrition and prolonged reduction in pumping can induce metabolic stress (USEPA, 1976). Significant reductions in pumping rates of adult oysters and clams occurred at silt concentrations of 100 mg/l. At higher concentrations, oysters ceased pumping altogether and prolonged exposure led to death (Loosanoff, 1961).

Effects on shellfish species during aggregate dredging and capital and maintenance dredging

Capital, maintenance and aggregate dredging

Again, many of the generic effects on shellfish associated with sediment plumes are similar for both aggregate dredging and capital and maintenance dredging. Therefore, a detailed description of the different effects arising is not necessary.

5.5.4 Effects on commercial shellfisheries

The impacts discussed above can affect commercial shellfisheries. For example, the potential accumulation of fine sediment in crab and lobster holes may cause a redistribution of these species. This can reduce the catch from traditional fishing grounds as a result of stock redistribution and silted pots (Emu Environmental, 1998).

The brown crab and scallop are known to spawn within an area from which it was proposed to dredge sand and gravel from Inner Owers (Emu Environmental, 1998). The potential for these stocks to be adversely affected was identified and put into the context of national resources. Although it was concluded that they represented an insignificant proportion of the total English Channel stock, the effect of stock reduction on local fishermen could be significant. It was noted that dredging had occurred on the Outer Owers for a considerable time, yet the area remained a productive fishery for scallop.

Box 5.4 *Effect of sediment plumes on berried hen crabs at Area 107, Race Bank, off eastern England, UK*

> The study was based on the deployment of autonomous bottom-landers, known as minipods. The minipods are platforms that carry timed sediment traps, passive sediment traps and an array of main sensors capable of measuring near-bottom suspended sediment. The aim of the study was to undertake autonomous, reliable appraisals of suspended sediment due to dredging at various ranges from the dredging activity at Area 107.
>
> A linear array of four minipods were set between the dredge site and Race Bank in May–June 1995, with the sensors recording at the maximum (near-continuous) data-rate. The first deployment of minipods as part of Contract A0902, in May–June 1995, showed that spikes in suspended load, supposedly due to dredging at Area 107, were only present at Race Bank during spring tides. During neap tides these peaks were absent (CEFAS, 1998). This was found to be related to the tidal excursion on spring and neap tides. The tidal excursion on spring tides was sufficiently broad (about 9 km) to span the 6.5 km distance from the dredge site to Race Bank, whereas the 4 km excursion during neap tides was not. The tides could therefore convey the outwash sediment from Area 107 to Race Bank during spring tides (CEFAS, 1998).
>
> The simultaneous deployment of four minipods between Area 107 and Race Bank aimed to provide information regarding the movement of sediment plumes between Area 107 and Race Bank. The experiment provided evidence that an individual outwash plume passed in sequence from one minipod to the next across the whole distance from the dredge site to Race Bank. The study concluded that dredging at Area 107 has the potential to deliver an extra 50–150 mg/l of suspended sediment to the near-bottom layer at Race Bank during about 7 per cent of the spring/neap cycle. Therefore interference with the crabs at Race Bank will be small (CEFAS, 1998).
>
> A less sensitive measure of the overall geographical extent of the sediment plumes was later obtained using downward-looking 1200 kHz acoustic profiler during February 1996. This method showed that the sediment plume could be distinguished from background concentrations at up to 2500 m range during the spring tide conditions but only 1200 m during neap tides (CEFAS, 1998).
>
> Between April and July 1997, five minipods were deployed over the wider area around Area 107, with deployments at Nut and Spanner (8.5 km distant from Area 107), Burnham Flats Buoy (16 km distant), Burnham Flats (18 km distant) and off Skegness (25 km distant). The results from these deployments showed that suspended sediment plumes from dredging activity reached as far as Nut and Spanner but did not reach the three most distant minipods at Burnham Flats Buoy, Burnham Flats or Skegness (CEFAS, 1998).

Field assessment of aggregate dredging effects on berried hen crabs

CEFAS has undertaken studies on the impact of sediment plumes derived from dredging on Area 107 off the east coast of England on the Race Bank and surrounding areas (CEFAS, 1998). This work was sponsored by MAFF under Contract A0902. The study was initiated to investigate concerns that dredging-derived sediment plumes from Area 107 may pose a significant risk to the berried hen crabs on Race Bank and consequently affect brown shrimps in the Wash. The study is the most comprehensive and scientifically advanced to date on the impacts of dredging on the movements of suspended sediments. The information about the effects of aggregate dredging in Box 5.4 (above) has been prepared for CIRIA RP600 by CEFAS with the approval of South Coast Shipping.

5.6 OTHER ENVIRONMENTAL EFFECTS

5.6.1 Seabed sedimentology

Seabed sediment types and morphologies

The seabed is formed of a variety of sediment lithologies, including cohesive and non-cohesive silt and clay, fine- to coarse-grained sand, gravel, cobbles and boulders, or of glacial till (cohesive sediment with a wide range in grain size) or bedrock. Its topography ranges from featureless (eg the lag gravel pavements that form much of the North Sea bed) to steeply banked (eg the accretionary sand ridges up to 30 m high known as the Norfolk Banks in the southern North Sea).

At a smaller scale, the seabed may be formed of accretionary sand deposits such as waves, megaripples or ribbons (Evans *et al*, 1998). The nature of the seabed is influenced largely by its sediment supply and the wave and tidal regimes to which it is subject. The extent to which sediment either accretes or is eroded and transported across the seabed is determined by those regimes. Wave-induced and tidal currents vary greatly in space and over time, and their capacity to erode, transport and deposit sediment varies accordingly. For example, in the UK, extreme wave events cause major morphological changes to the Yarmouth Banks (sand) off East Anglia.

Consequences of plume settlement on seabed sedimentology

The settlement of a sediment plume on the seabed can alter the sedimentology of seabed areas beyond the dredged area. In particular, the screened material released during aggregate dredging has the potential to affect sedimentological conditions by returning a disproportionate amount of fine or coarse sediment to the seabed. Rates of return vary, but Tables 5.4 and 5.5 from Newell *et al*, 1998 (based on Hitchcock and Drucker, 1996) show, respectively, the size distribution of overspill and reject material and the screened load quantities from a trailer suction hopper dredger during aggregate dredging.

Table 5.4 *Particle size distributions of overspill and reject material (from Newell et al, 1998; based on Hitchcock and Drucker, 1996)*

Particle size (mm)	Proportion of spillway discharge (%)	Proportion of reject chute discharge (%)
<0.063	38.0	1.0
0.063–0.125	14.0	0.9
0.125–0.250	5.7	8.9
0.250–0.500	12.9	31.4
0.500–1.000	9.2	27.3
1.000–2.000	3.3	12.0
>2.000	16.9	18.5

Table 5.5 *Screened load quantities (from Newell et al, 1998; based on Hitchcock and Drucker, 1996)*

	Dry solids (tonnes)	Water (tonnes±5%)
Quantity pumped	12 158	33 356
Quantity retained	4185	874
Quantity rejected through screening	7223	13 499
Quantity lost through overspill	750	21 387
Total losses	7973	34 886

The immediate physical consequences of the delivery of sediment to the seabed as a plume footprint depend on the wave and tidal regimes at the place and time of delivery. Considering an extreme condition of no current, sediment from the plume of any grain size could be expected to accrete, whatever the nature of the substratum. Thus a pre-existing sand deposit might become draped with a lamina or layer of silt or clay, for example. At the other extreme, if the plume footprint were to impact in peak spring tidal conditions, accretion of any sediment might be unlikely. If accretion of sediment does occur, it may be only temporary. As tidal currents increase within their diurnal or seasonal cycles, or as storm events generate strong, wave-induced currents, so plume-derived accretionary sediment may be eroded and transported away from its initial site of deposition. In the case of silts and clays, such reworking may be in the form of re-suspended fines or in a cohesive form as pellets and flakes that may subsequently become incorporated into sand deposits.

In summary, impact assessments considering the effect of the sediment plume should not neglect the potential effect on the nature of the seabed itself. In assessing the impacts of plumes on seabed sediments, it is important to have baseline knowledge of the sediment lithologies and morphologies occurring in the area under review, as well as information on the wave and tidal current regimes, and their likely variation over time. This information provides a perspective of the physical substratum, and likely changes to that substrate, in the assessment of the effects of plumes on the benthic ecology. In addition, knowledge of the variability of sedimentary conditions over time is important in any consideration of the use of environmental windows in dredging operations.

5.6.2
Capital, maintenance and aggregate dredging

Nature conservation sites

The potential effects that sediment plumes can have on marine ecology are described in Section 5.3. However, they can be more significant if the plumes affect a site designated for its nature conservation value. Examples of such sites in the UK include nationally important sites (eg National Nature Reserves and Sites of Special Scientific Interest) and internationally important European conservation sites (such as SPAs and SACs). In the UK, these sites are protected by legislation, including the Conservation (Natural Habitats etc) Regulations 1994 (see Appendix 1).

Sites are notified for their nature conservation importance either because they have a particularly diverse assemblage of species or because they contain rare or uncommon species. The designation requires more consideration to be given to the potential impact of a sediment plume (or other disturbance) on the integrity of the conservation resource.

The potential for dredging to affect European sites of nature conservation importance is under investigation in the UK. For example, the UK Marine SACs Project involves all of the UK's statutory nature conservation agencies and aims to facilitate the establishment of management schemes for 12 of the UK's candidate Marine SACs. One of the main components of this project is to assess the interactions that can take place between human activities and the habitat and species interests at these sites. Of the seven projects undertaken, one specifically investigates aggregate dredging and includes a review of the effects of suspended sediment.

5.6.3
Capital, maintenance and aggregate dredging

Recreation

Capital and maintenance and aggregate dredging have the potential to conflict with recreational activities such as fishing and diving. The following potential effects on recreation were identified in studies into aggregate dredging in the UK (but could equally apply to capital and maintenance dredging).

The potential for sediment plumes to impact upon recreational users (sport fishing) was highlighted during the examination of a proposal for aggregate dredging on Helwick Bank in the Bristol Channel (Gibb, 1997). Gibb concluded that the plume would not cause any significant discoloration of the water at sites remote from Helwick Bank and that the density of the plume was likely to be reduced to background levels within a short distance of its origin. It was noted that the disturbance caused by dredging would temporarily preclude sport-fishing activity in the immediate vicinity of the dredged area.

ERM (1999) considered that the extraction of sand from the Middle Bank, Firth of Forth could have the potential to impact recreational beaches situated within the Firth of Forth due to the deposition of fines from sediment plumes. Even under worst-case conditions, however, ERM suggested that dredging-induced sediment plumes were unlikely to have any significant impact on bathing waters and amenity beaches in the area (ERM, 1999).

5.6.4
Capital and maintenance dredging

Archaeology

The deposition of sediment from a plume can benefit archaeological resources. A benefit can arise if a site or wreck receives additional sediment cover, thereby protecting it from erosion by currents. Additionally, increased sediment cover could reduce the amount of oxygen reaching the archaeological remain, thus reducing the rate of degradation of organic materials. Since most marine archaeology is situated closer to land, it is more likely to be affected by sediment plumes arising from capital and maintenance dredging.

5.6.5 **Visual impact**

Capital, maintenance and aggregate dredging

The visual impact associated with a sediment plume is linked to the change in the aesthetic quality of the water. In most cases, the water's aesthetic quality is changed in terms of water clarity and/or colour.

The degree of an impact depends on the subjective opinion of the receptor affected by the visual change caused by a sediment plume. An opinion may be influenced by such factors as the ambient turbidity of the water and the environmental setting in which a sediment plume is created. For example, it is easier to appreciate the visual impact of a dense sediment plume in normally clear tropical waters than of a sediment plume that is barely distinguishable in dark high-energy waters such as the inner Bristol Channel.

5.6.6 **Beneficial effects**

Not all of the environmental effects of sediment plumes are negative. In some cases the introduction of sediment into the environment can indirectly cause beneficial effects. Minor benefits include providing a temporary food source for sea birds and fish due to the release of organic matter. More significantly, and of recent interest in the UK, is the use of suspended sediment to offset erosion of intertidal areas.

The intertidal areas of some UK estuaries, particularly in eastern England, are slowly eroding due to sea level rise, coastal squeeze and other factors. Inter-tidal mudflats and other habitats support benthic faunal species, which vary according to substrate type. In turn, the benthic fauna provides a food resource for nationally and internationally important numbers of over-wintering waterfowl and wading birds.

In some cases the gradual erosion of estuarine intertidal areas can be reduced or temporarily offset by introducing additional sediment into the system, thereby benefiting the local ecology. This approach is based on the understanding that the long-term evolution of an estuary's intertidal areas involves a balance between periods of sediment erosion and deposition. Accordingly, if the amount of deposition can be increased, then the balance can be moved towards greater net accretion or reduced net erosion.

This approach was taken as a partial mitigation measure to offset the impact of erosion caused by the capital dredging of the approach channel to Harwich Haven and the Port of Felixstowe. It was suggested that dredgers could be used to increase the suspended sediment concentration in the Stour and Orwell estuaries by direct water column recharge and increased overflowing during maintenance dredging. In both cases, the methodology for the introduction of sediment was designed so that:

- the increased suspended sediment concentration must be effective when the water lies over the intertidal areas at high water
- the sediment introduced should increase the suspended sediment concentration by a small amount over as wide an area as possible and for as long as possible to promote a steady, natural mode of accretion rather than rapid patchy accretion, and to avoid excessive turbidity
- smothering of the seabed by thick sediment deposits is minimised
- the introduction of sediment can be effected in a practical way (HR Wallingford and Posford Duvivier Environment, 1998).

5.7 CUMULATIVE EFFECTS

5.7.1 Introduction

Cumulative effects arise as a consequence of more than one dredging operation affecting a defined zone of influence. Cumulative effect assessment recognises that the combination of repeated dredging at one site, or more than one dredging operation at nearby sites, increases the pressure on natural resources normally associated with a single dredging regime. In isolation, one dredging regime might not significantly impact upon, for example, a fishery area; however, the cumulative effect of several dredging regimes might be to exceed to a critical threshold, preventing, for example, the recovery of the fishery.

In the context of sediment plumes a cumulative impact is considered to constitute more than one dredging activity creating multiple plumes, although there also could be interactions between the plume(s) and other dredging effects.

The Canadian Environmental Assessment Agency (1999) defines a cumulative (or combined) environmental effect as "the effect on the environment that results from the effects of a project combined with those of other past, existing and imminent projects and activities. These may occur over a certain period of time and distance" – ie the interaction between and combined influence of all works proposed or planned within the zone of influence of the project in question.

Cumulative effects can be additive or interactive. Additive effects occur when one unit of environmental change in one area adds to or subtracts from the environment in another area. Interactive effects occur when the net accumulation of environmental change is more or less than the sum of all environmental changes.

The assessment of the environmental effects associated with a proposed dredging project is usually undertaken as part of a licence or consent application procedure and in accordance with statutory legislation on EIA. Until recently, EIAs undertaken in the UK were not explicitly required to assess the cumulative effect of more than one (dredging) project on the environment. However this situation has changed since the introduction of the Environmental Impact Assessment and Habitats (Extraction of Minerals by Marine Dredging) Regulations 2000, with respect to aggregate dredging, and the Conservation (Natural Habitats etc) Regulations 1994, with respect to capital and maintenance dredging. Both require cumulative impacts to be considered (see Appendix 1).

5.7.2 Cumulative environmental impact assessment

A thorough cumulative environmental impact assessment (CEIA) for a project-specific investigation of a proposed aggregate or capital and maintenance dredging campaign should consider the cumulative effects of other activities in conjunction with the dredging (including other dredging). The impacts associated with non-dredging activities are likely to be wider in scope than the potential impact of sediment plume arising from the dredging itself.

Although (until recently) the majority of EIAs do not undertake to address cumulative effects explicitly, any EIA of a proposed dredging project essentially considers the cumulative effects of the proposed project in addition to past and existing projects. This is, because the environmental baseline against which the EIA is carried out is itself affected by past and existing projects. For example, the heavily modified condition of the San Francisco Bay estuary is a result of activities regulated by a wide variety of government agencies (National Performance Review, 1994 in CEQ, 1997).

However, CEIA also requires the consideration of the environmental effects of the proposed project with those of future projects, in addition to past projects. In Canada, for example, future projects include (at a minimum) those that have been approved under the Canadian Environmental Assessment Act. The environmental effects of uncertain or hypothetical projects need not be considered.

CEIA in the United States

In the United States, the Council on Environmental Quality (CEQ) has produced guidelines on the consideration of cumulative effects under the National Environmental Policy Act, 1969. The guidelines discuss the incorporation of cumulative effects analysis (CEA) into the following components of environmental impact assessment:

- scoping
- describing the affected environment
- determining the environmental consequences.

The steps and general principles involved in this process are summarised in Table 5.6.

Table 5.6 *The US CEA process (CEQ, 1997)*

EIA component	CEA steps	Principles
Scoping	**Step 1:** identify the significant cumulative effects associated with the proposed action and define the assessment goals	Include past, present and future actions
	Step 2: establish the geographic scope for the analysis	Include all federal, non-federal and private actions
	Step 3: establish the time frame for the analysis	Focus on each affected resource, ecosystem and human community
	Step 4: identify other actions affecting the resources, ecosystems and human communities of concern	Focus on truly meaningful effects
Describing the affected environment	**Step 5:** characterise the resources, ecosystems and human communities identified during scoping in terms of their response to change and capacity to withstand stresses	Use natural boundaries Focus on each affected resource, ecosystem and human community
	Step 6: characterise the stresses affecting these resources, ecosystems and human communities and their relation to regulatory thresholds	
	Step 7: define a baseline condition for the resources, ecosystems and human communities	
Determining the environmental consequences	**Step 8:** identify the important cause-and-effect relationships between human activities and resources, ecosystems and human communities	Address additive, countervailing and synergistic effects
	Step 9: determine the magnitude and significance of cumulative effects	Look beyond the life of the action
	Step 10: modify or add alternatives to avoid, minimise or mitigate significant cumulative effects	Address the sustainability of resources, ecosystems and human communities
	Step 11: monitor the cumulative effects of the selected alternative and adapt management	

UK investigations into the cumulative effects of aggregate dredging

CEFAS is undertaking a project concerned with the cumulative environmental impacts of marine aggregate extraction. The project's scope is not confined to the effects of sediment plumes, but is concerned with all potential cumulative impacts associated with marine aggregate extraction (Box 5.5 gives more details of the aims of the research).

CEFAS has provided the rationale for addressing cumulative impacts. As a starting position, it is known that, at an isolated area licensed for dredging, impacts occur as a direct consequence of passage of the draghead. It is also known that there is scope for indirect effects both within and beyond the dredging area, especially arising from the dispersion and settlement of fine sediment, although actual evidence of its extent and severity is limited.

Even in isolation, the cumulative consequences of dredging could be locally severe (eg by irreversibly changing the prevailing substrate type or by exceeding the medium-term capacity of the locality to disperse fine sediment disturbed or discharged during dredging). Because such activity is typically conducted on a small geographical scale, however, it is hard to conceive of circumstances where such activity could, for example, impinge significantly upon the stock size of fish species of commercial interest.

CEFAS suggests that cumulative effects of a sort may occur as a result of dredging at a single locality but, in general, studies to identify such effects will be insufficiently wide in scope to account for current pre-occupations with the cumulative consequences of aggregations of licensed dredging activity. In this context, cumulative impacts of ongoing dredging activity are perceived to occur on geographical scales that are sufficient to affect stock sizes and hence fishermen's livelihoods.

Box 5.5 *CEFAS research project – cumulative environmental impacts of marine aggregate extraction*

Aims

1. To identify the scope for cumulative effects of multiple dredging activity on the seabed environment and fisheries, initially with special reference to licensed areas off Lowestoft and the Isle of Wight.

2. To describe the current environmental status of these regions, with particular attention to the biological communities and the factors controlling them.

3. To plan and execute a programme of scientific sampling of key biological and environmental measures relevant to the assessment of cumulative impacts, including definitions of the appropriate spatial scales for benthic sampling in order to identify any biological effects.

4. To quantify the spatial scale of physical and biological impact arising from aggregate dredging within selected localities, and its potential significance, especially for commercial fisheries, and variability with time.

5. To develop a predictive capability, allowing the consequences of any future extension of dredging activity to be evaluated at the licensing stage.

6. To help to define the circumstances under which cumulative effects studies may become necessary in other areas, and to develop a generic framework for such studies.

7. To report on findings, with particular emphasis on information beneficial to the assessment of licence applications.

To the extent that such cumulative consequences might arise, the concern is that dredging might reach a critical intensity beyond which a fishery population may be reduced to sub-commercial levels. CEFAS suggests there is no basis for presuming a simple additive relationship between dredging activity and the size of stock, nor are there adequate data to test for such a relationship. Accordingly, any demands for increases in the quantities to be dredged, either from within existing clusters of licensed activity or from newly licensed areas nearby, are likely to accentuate this concern.

The following paragraphs describe CEFAS's approach to investigating cumulative effects of aggregate dredging.

In an initial assessment of the scope for cumulative effects, CEFAS considers that a review of historical data will be very important, particularly with regard to dredging intensity and extent, and the relative performance of local fisheries. However, the initial assessment will not provide definitive answers because of difficulties in obtaining reliable data of sufficient resolution to quantify adequately any effects within the precise local geographical areas of concern. CEFAS will therefore also place strong emphasis on the conduct of new, carefully targeted sampling regimes to cover appropriate spatial scales, and especially to establish the stability of observed effects over time.

This temporal (year-on-year) element to CEFAS's sampling will be critical to establishing with confidence that real changes are evident, and that the changes are not simply an artefact of sampling on one occasion only. Factors such as annual variability in dredging intensity, weather conditions affecting local fishing effort, natural variability in fish/shellfish stocks (eg shrimp surveys in the Wash) and in benthic populations determine that one-off sampling is inadequate to address this issue effectively.

The challenge of CEFAS's project is to distinguish natural from dredging-induced changes, thereby allowing an objective scientific evaluation to be made of the continued acceptability of multiple extraction activities.

5.7.3 Measuring cumulative effects

Little work has been done on measuring the cumulative effects of multiple sediment plumes in the UK, since the need to address them has only been introduced recently. CEFAS has undertaken the most advanced work so far, using minipods, as described below.

The use of minipods in the study of sediment plumes and their movement in the region of Area 107 and Race Bank, off the east coast of England, has provided evidence of the potential for cumulative effect of dredging with other natural and anthropogenic phenomena.

Before the minipod deployment in May–July 1995, minipods were deployed initially in the winter of 1994–1995, at a time when no dredging activity was taking place in Area 107. This deployment showed the relative amplitudes and durations of a major storm and a major flood event. The minipods recorded several individual and short-lived peaks in suspended loads at the time of a storm around 1–3 January 1995. This local effect was overwhelmed several days later when the floodwaters, caused by the same storm, ran off eastern England and spread offshore as a turbid and greatly extended plume from the River Humber. This flood maintained a high turbidity at the Race Bank more or less until the end of the record on 18 January 1995, long after the storm that gave rise to it had passed (CEFAS, 1998).

This observation highlights the potential for natural phenomena to cause suspended sediment increases. The study concluded that suspended loads experienced in winter at Race Bank are typically much higher than arise in the summer due to the spreading of outwash material from Area 107.

The results from the deployment of minipods in the wider area around Area 107 showed the potential for cumulative impact of other anthropogenic activities with aggregate dredging. The minipod deployed at Burnham Flats Buoy recorded abnormally large spikes in suspended load that were uncorrelated with dredging activity at Area 107. Additionally, the peaks were most frequent during neap tides during calm weather when the processes of local re-suspension due to wind and tide are minimal. It was concluded that disturbance of the seabed due to beam trawlers working the edge of Burnham Flats and Ridge was the most likely cause of the elevated suspended sediment. This conclusion would, however, need to be correlated with fishing records (CEFAS, 1998).

Future use could be made of this technique for monitoring, but it must be stressed that a good baseline is essential. Monitoring of other activities occurring, both natural and anthropogenic, is required in order to explain possible anomalies in results.

6 Mitigating environmental effects

This chapter reviews recent work concerned with options for mitigating the extent of sediment plumes. It focuses on two fundamental areas of mitigation:

- choice and operation of dredging plant, with examples from MMS (1996)
- environmental windows, with examples from Reine *et al* (1998) and CEFAS (1998).

6.1 CHOICE AND OPERATION OF DREDGING PLANT

The choice and operation of dredging plant with respect to the environment to be dredged is fundamental to reducing sediment plumes arising from dredging. Section 3.2 provides descriptions of various common dredging plant and details the principal causes of sediment re-suspension associated with each type of dredger.

Table 6.1 summaries the measures by which the dredger and its operation can be improved in terms of reducing sediment losses to the water column. No summary is given for hydrodynamic dredgers, since sediment re-suspension is a critical part of the dredging process, or for environmental dredgers since the latter are often designed to minimise sediment re-suspension.

Table 6.1 *Measures to reduce sediment re-suspension from dredgers*

Dredger	Mitigation measure
Trailer suction hopper dredger	Optimise trailing velocity, suction mouth and pump discharge Limit overflow and/or hopper filling Reduce intake water Use return flow Reduce air content in the overflow mixture
Cutter suction dredger	Optimise cutter speed, swing velocity and suction discharge Shield the cutter head or suction head Optimise cutter head design
Grab dredger	Use a hydraulic, watertight grab Use a silt screen (discussion follows) Limit grab time above water Limit dragging on the seabed
Backhoe dredger and dipper dredger	Use a visor over the bucket Use a silt screen (discussion follows)
Bucket ladder dredger	Optimise bucket filling, slack in bucket chain, swing control, rates of bucket filling and dredger advance Maintenance of bucket chute Install splash screens Use one-way valves in buckets Enclose ascending part of the bucket ladder

6.1.1 Trailing suction hopper dredging

Limiting sediment re-suspension

Sediment re-suspension caused by trailing suction hopper dredging can be reduced in the following ways:

- trailing velocity, position of the suction mouth and the discharge of the pump can be optimised with respect to each other
- any reduction in the intake of water by the suction head means a more dense pay load and thus reduces the need for overflowing. This can be achieved by directing the flow lines of the suction stream to the actual point of excavation thus making better use of the erosive capacities of the flow of water into the suction head
- under certain circumstances the "return flow" method offers a possible improvement for the trailing suction hopper dredging process (Van Doorn, 1988). With this method, the water in the hopper (which would otherwise be discharged overboard or transported to the disposal site) is returned to the suction head to contribute once again to the erosive flow. This limits the total intake of water and increases the pay load
- a lot of material from the overflow is suspended at the surface because of the presence of air in the overflow mixture. Much of this air can readily be taken out onboard by installation of a well-designed overflow system. As a density flow, without air, the overflow mixture descends more quickly to the seabed.

Overflow mitigation measures

Several examples of mitigation measures are described for marine aggregate mining in Massachusetts Bay using a trailing suction hopper dredger (MMS, 1996). Some of these measures were designed to mitigate adverse impacts on water quality and comprised mitigating overflow from the dredger (MMS, 1996). Brief descriptions of the following MMS overflow mitigation measures are provided below:

- operational technique
- recycling
- hopper design
- overflow collection
- effluent discharge
- operational mitigation
- design mitigation.

A practical operational mitigation technique is to delay the commencement of the dredge hopper overflow. This can be achieved by emptying all water out of the hopper before starting to pump any sediment on board. Water can be discharged directly overboard as the dredge pump is primed. A switching mechanism can then be used to direct the slurry to the hopper once sediments are entrained in the pumping system. As a result, overflow cannot occur until the hopper is filled to within 60–70 per cent of its dredged material capacity. The time during which sediment-laden waters are released to the sea can therefore be substantially reduced. It should be noted that 60–70 per cent is the maximum fill point to which overflow can be delayed and assumes the overflow (or weirs) is set at its highest level. Even then, a 60–70 per cent filling before overflow is only likely when dredging mud or very loose fine sands. In other materials, the filling capacity may be much lower.

The drag-heads of some trailing suction hopper dredgers are equipped with water jets designed to assist in the liberation of compacted sediments when dredging hard ground. A small percentage of the hopper overflow water can be filtered to create clean water, which is then pumped to the drag-head jets, recycling it instead of using clean seawater. This measure can slightly reduce the volume of sediment-laden overflow.

The level of the aggregate surface in a loaded dredge hopper always must be above the load line in order to prevent surplus water from being transported to shore in addition to the solids. This can be achieved by modifying the designed shape of the hopper to keep much of the contents above the load line.

A simple technique for handling hopper overflow, called an anti-turbidity overflow system (ATOS), has been developed in Japan. The overflow collection system is streamlined to minimise the entrapment of air bubbles in the overflow water. Removal of air bubbles, which otherwise makes particles more buoyant and prolongs settling, allows the fine particles to settle at a faster rate.

6.1.2 Cutter suction dredging

Limiting sediment re-suspension

Sediment re-suspension caused by cutter suction dredging can be reduced in the following ways:

- optimising the cutter speed, swing velocity and suction discharge with respect to each other
- putting a moveable shield around and above the cutter head or suction head reduces the escape of suspended sediment into the surrounding water column
- optimising the design of the cutter head with respect to the material being dredged to improve the movement of sediment toward the suction intake.

Sediment re-suspension from cutter suction dredging is greatly affected by the swing and rotation of the cutter head; this is born out by laboratory research (see Appendix 3). Limiting these two variables can theoretically halve sediment losses from dredging, although this rate is not always possible in practice.

Mitigating water quality impacts

Other measures have been designed to mitigate water quality impacts and have been applied to the dredging of heavy minerals offshore of Virginia Beach, Virginia, using a cutter suction dredge. The MMS (1996) refers to such measures as:

- operational mitigation
- design mitigation.

The MMS recommends certain operation mitigation controls to reduce sediment re-suspension. Most involve the manner of excavation. For example, very deep single-pass cuts should be avoided because the cutter-head tends to become buried. As a result, more sediment is excavated and dislodged than the suction can entrain, leading to high levels of suspension. A thick deposit is best excavated by means of several horizontal slices, each layer of excavation being the depth of a single cut.

The suction at the cutter head should be sufficiently powerful to collect all disturbed sediment. Pick-up capability could be increased by including water jet booster systems or ladder-mounted submerged pumps. Water jet boosters generally increase sediment re-suspension so they need to be set up to direct sediment towards the cutterhead. The cutter head may be designed in such a way that the suction is brought closer to the sediments thereby improving the chances of entrainment.

Proper design of the cutter with an appropriate rake angle based on nominal cutter revolutions (ie the acute angle between the base of the cutter tooth and the upper surface of the sediment) is an important mitigating factor. If the rake angle is too large, a gouging action can throw soft, fine-grained sediments outward. If the rake angle is too small, however, re-suspension can occur on impact with the bed.

6.1.3 Grab dredging

Sediment re-suspension caused by grab dredging can be reduced in the following ways:

- using a watertight grab
- using a hydraulic grab
- using a silt screen (see Section 6.1.7)
- limiting the swinging of the grab over open water, thereby reducing the time when soil can leak out of the grab
- limiting the practice of smoothing the excavated area by dragging the grab along the bottom.

A watertight grab is closed from above by rubber flaps or by steel plates and reduces the release of the fines-water mixture when lifted through and above the water. Alternatively, a hydraulic grab enables the controlling and monitoring of the opening and closing actions of the grab. If the closure process is prevented by an obstruction, this can be dropped just above the bottom and the grab process can be restarted.

The losses caused by grab dredging can be reduced by up to 50 per cent by making the grab watertight (closed) and can be reduced by up to a further 50 per cent by the use of a silt screen. These reductions are born out by the field data presented in Appendix 3.

As the size of grab increases the absolute loss of sediment increases. However, the production rate increases more rapidly with the grab size than the loss rate and so the total loss per unit production (the S-factor) decreases with bucket size.

6.1.4 Backhoe dredging

Sediment re-suspension caused by backhoe dredging can be reduced in the following ways:

- using a silt screen (see Section 6.1.7)
- using a visor grab (similar to a closed grab for grab dredging)
- limiting the swinging of the backhoe over open water, thereby reducing the time when soil can leak out of the bucket
- limiting the practice of smoothing the excavated area by dragging the backhoe bucket along the bottom.

6.1.5 Bucket ladder dredging

Sediment re-suspension caused by bucket ladder dredging can be reduced in the following ways:

- optimising the degree of filling of the buckets, the amount of slack in the bucket chain, controlling the swing, the advance of the dredger, the rate of bucket filling and the bank height
- good maintenance of the discharge chutes prevents leakage; further leakage can be prevented by stationing a barge under the chute that is not being used for loading, although this is an expensive measure
- installing splash screens at the end of the chutes to further reduce loss of material during the loading process
- inserting one-way valves in the bottom of the buckets
- enclosing the ascending part of the bucket ladder.

The insertion of one-way valves in the bottom of the buckets enables the extraction of air from the buckets as they descend through the water column and also helps the emptying of buckets when dredging soft cohesive sediment. Enclosing the ascending part of the bucket ladder reduces the ability of currents to wash away material from the buckets and causes any material falling or being washed out of the buckets to quickly fall to the bed as a dense mixture. As for grab and backhoe dredging, the loss of sediment increases with the size of the bucket while the total loss per unit production (the S-factor) decreases with bucket size.

6.1.6 Environmental dredging

Environmental dredging focuses either on operating plant with minimal re-suspension or with particular accuracy, thereby limiting sediment re-suspension. The operation and performance of environmental dredgers is described in Section 3.2.12.

6.1.7 Silt screens

The release of sediment from some stationary dredgers (such as grab and backhoe dredgers) can, under certain conditions, be reduced by use of a silt screen. A silt screen is a curtain of cloth suspended from a floating framework down to the bed. The cloth must be permeable to water but not to silt. Generally, the screen aims to reduce the spreading of fine suspended sediment through the flow vertical. By this means the area in which the re-suspension occurs is isolated from the surrounding water. When the area within the silt screen has been dredged to the required depth, the suspended sediment will settle locally. Considerable attention must be paid to the movement of the screen as careless handling of the screen may completely negate the advantages of its use.

The effectiveness of silt screens is limited, however, to areas where the ambient currents are small and wave conditions are low. Situations with tidal currents in excess of 0.5m/s are unsuitable for silt screens. Hence, the use of silt screens is not always practical or economic, and expert attention is needed during their design and use. The additional costs and problems presented by the use of silt screens means that other environmental measures can frequently be more suitable.

The MMS (1996) identified the following limitations to the use of silt screens:

- they cannot be used where currents are more than 0.5 m/s
- waves and swell render silt curtains inoperative, so they can only be used in calm conditions such as in harbours and estuaries
- they are only useful for small-scale, anchored operations
- the manoeuvrability of the dredge is limited
- the movement of the screen after excavation of the area inside can cause the explosive release of sediment.

Given the above, the potential for the effective use of silt screens for controlling sediment plumes is very limited.

6.2 ENVIRONMENTAL WINDOWS

Environmental windows are essentially time zones that provide mitigation by restricting dredging to avoid adverse environmental effects. Accordingly, the USACE defines environmental windows as "specified time periods to which a dredging project should be confined". The aim is to protect sensitive biological resources or their habitats from the potentially detrimental effects of dredging by avoiding dredging-induced perturbations while resources perceived to be at risk are present (Reine et al, 1998). Environmental windows are typically incorporated into dredging licence conditions.

In the USA, frequent use is made of environmental windows in an attempt to mitigate for the impacts associated with capital (or operations) and maintenance dredging. Their use in the UK is increasing, although restricted by a lack of knowledge about the site-specific factors needed to identify appropriate windows. In terms of sediment plumes, there are two fundamental sets of site-specific factors to take into account:

- the environmental interest(s) at risk from the re-suspended sediment plume or its subsequent settlement on the seabed (eg location of shellfish beds, fish migration routes and seasons)
- the environmental factors affecting the distribution and impact of the re-suspended sediment plume and its subsequent settlement (eg tidal excursion, sediment type and quantity).

6.2.1 US experience of environmental windows

In the USA, about 80 per cent of all capital works and maintenance dredging projects are subject to environmental windows. Requests for environmental windows from various agencies relate to a range of environmental issues. In a survey of environmental windows issues, that was sent to 38 USACE districts, the following were cited as causes for requests for environmental windows:

- disruption of avian nesting activities and destruction of bird habitat
- sedimentation and turbidity issues involving fish and shellfish spawning
- disruption of anadromous fish migrations
- entrainment of juvenile and larval fishes
- entrainment of threatened and endangered sea turtles as well as disruption of their nesting activities during beach nourishment projects
- burial and physical removal of protected plants
- disruption of recreational activities.

Of the USACE districts surveyed, 68 per cent reported turbidity, suspended sediments, and/or sedimentation as a reason for environmental windows (Reine *et al*, 1998). Table 6.2 summarises the findings of the USACE survey on environmental windows regarding the effects associated with sediment plumes.

Most applied environmental windows constrain dredging operations during spring and summer in the USA (March–September) to avoid potential conflicts with biological activities such as migration, spawning and nesting. Consequently, many dredging projects must occur in the winter (when adverse weather conditions can, in fact, cause more dispersion of the plume and increase the likelihood of spills). Dredging during winter also has potentially more downtime, and thus can become more expensive.

However, many USACE districts receive requests for restrictions in all seasons, which results in difficulties in fulfilling dredging requirements. Cases have arisen where there has been a cumulative impact of multiple restrictions applying to the same dredging operation. The result of this has been that dredging activity in many navigable waterways can be curtailed throughout much of the year (Reine *et al*, 1998).

Table 6.2   *USACE environmental windows survey regarding sediment plumes*

Dredging-related impact	Potentially impacted environmental resource	Nature of impact	USACE districts expressing the impact as an issue of concern
Turbidity/ re-suspended sediment	Commercial and sport fisheries	Migration blockage Reduced water quality	59
Sedimentation	Commercial and sport fisheries	Smothering of spawning/nursery ground	41
Sedimentation	Aquatic plants (eg eelgrass)	Burial	14
Turbidity and sedimentation	Shellfish	Burial Smothering of spawning/ nursery ground Reduced water quality	24

Environmental windows are often viewed as being overly conservative and based largely on limited, poorly quantified data or on subjective opinion. For example, the potential blockage of migratory pathways of various anadromous adult and juvenile fish due to their theoretical avoidance of turbidity plumes was frequently an issue of concern associated with windows compliance by many USACE districts, however, such blockage has not been conclusively demonstrated by field studies (Reine *et al*, 1998). Therefore, in many cases it appears that a precautionary approach is being adopted.

Although environmental windows are generally complied with, this compliance is viewed as being problematic because it complicates scheduling, causes contractual delays and substantially increases project costs. In addition, by confining the activity to the winter months, there is an increase in the associated risks, such as personnel safety for dredge crew members and a tendency to limit contingencies for repairs and severe weather shutdowns (Reine *et al*, 1998).

The US experience of environmental windows has also raised the issue of attendant economic impacts, typically expressed in incremental costs per cubic metre of dredged material. The implementation of environmental windows as a mitigation measure can be a significant cost factor for dredging operations.

6.2.2 UK experience of environmental windows

Research by CEFAS into the movements of sediment plumes generated by aggregate dredging activity at Area 107 led to the use of environmental windows as a mitigation technique. Results of the minipod deployments showed the circumstances under which sediment plumes from dredging activities may reach the Race Bank over-wintering ground for berried hen crabs. The minipods proved that the transport of fine suspended sediment as far as Race Bank occurred only during spring tide conditions. It was therefore proposed that aggregate dredging of Area 107 was carried out only on a flood tide to carry the sediment plumes away from Race Bank. This practice was adopted by the dredging company (CEFAS, 1998).

Environmental windows have also been employed to mitigate the effects of dredging (including any plume) on inshore sole migration at the Hastings Shingle Bank dredging area. To minimise the potential for an effect, the licensees must cease dredging at night for an annual period of ten weeks (corresponding to the spring sole-fishing season). The period starts between mid-February and mid-March each year and the precise start date must be notified by the Hastings Fishermen's Protection Society, giving seven days notice (from licence conditions, 1st edition, 1996). The reason for this measure is that sole are believed to move mainly at night (CEFAS, pers comm).

7 Towards an assessment framework

7.1 THE REQUIREMENTS OF AN ASSESSMENT FRAMEWORK

This section outlines the components of a structured framework for assessing the environmental effects arising from dredging plumes. That is, it considers the steps needed to inform the decision-making process and the information and actions required to inform the steps. Such a framework must be sufficiently flexible to accommodate the different types of environment that characterise the maritime zone and the types of dredger that operate within it.

Figure 7.1 sets out the indicative components of such a framework for assessing the sediment plumes arising from dredging; regardless of the purpose for which the dredging is being undertaken, that is, aggregate, capital or maintenance. It should not be taken to represent the definitive framework, but simply the first step towards it (in line with the scoping objective of the report).

The assessment framework has been used to identify gaps in our current knowledge and the requirements for further work described in Section 8.

7.1.1 Description of the framework

In considering the framework it is important to realise that the process is not entirely sequential and that some activities and decisions will take place in parallel with others. However, it is recommended that a tiered or reiterative approach is adopted in order to both represent the complex nature of the potential interactions and to avoid spending time and money on assessing projects where it quickly becomes apparent that the environmental effects will be unacceptable, without having to resort to detailed monitoring or model studies.

Step 1 – Describing the dredging project

In assessing the sediment plumes arising from dredging, it is important to recognise that three main types of dredging occur – aggregate dredging, capital dredging and maintenance dredging – involving different plant, dredging different bed materials, typically operating in different environments. The description of the project must therefore include a broad assessment of the characteristics of the location (eg open coast or estuarine), the amount and type of material to be dredged and the type of plant that is expected to be used. This should provide enough information to decide whether or not a plume is likely to be generated.

Step 2 – Defining the nature of the plume

The plume generated will be dynamic or passive, or both. Dynamic plumes also generate passive plumes. The primary factors that determine the nature of the plume are the dredging technique, the sediment type, the hydrodynamic conditions and their interaction. The dredging technique will have already been determined (in Step 1) largely based on commercial criteria and technical acceptability. The sediment type

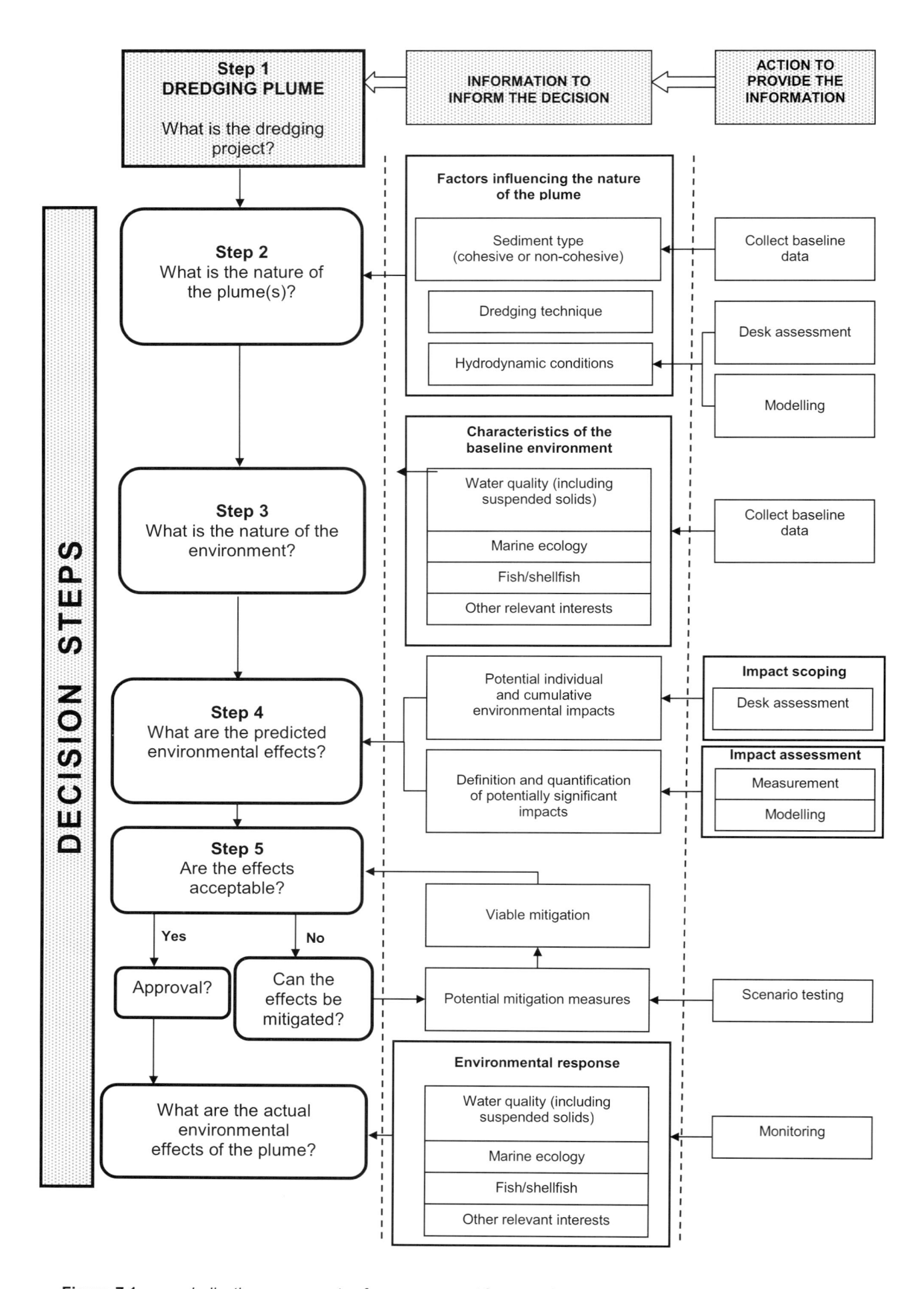

Figure 7.1   *Indicative components of an assessment framework*

must be known and this may involve field sampling, remembering that dredging involves the removal of material from below normal bed level and therefore that superficial sampling will not be sufficient. Whether the material is coarse and non-cohesive or fine and cohesive will have a significant bearing on whether a passive plume is likely to form.

The hydrodynamic conditions should be considered in terms of primarily currents, water depth and waves. Currents affect the degree of turbulence and the rate of advection. Water depth (in certain types of operation) affects the amount of sediment released. It also affects the type of plant likely to be used, but this is taken to have been predetermined in Step 1. Waves also affect turbulence. Their main effects are to inhibit settlement on the bed and to cause re-suspension. Describing the hydrodynamic conditions in the first instance may be achieved by reference to published information. For a more detailed assessment it is probable that modelling will be required. The need for this will probably depend on the sensitivity of the environment (Step 3).

The requirement for modelling may be limited to providing a description of the hydrodynamic environment. However, if more detail is required then plume modelling will probably be required. While passive plume modelling can be undertaken with reasonable confidence in the results concerning the extent of influence of the plume, the definition of the source term (ie the amount and rate of sediment release) is not well defined or predictable (see Section 7.2).

Step 3 – Defining the nature of the environment

To predict and assess the effects of dredging-induced sediment plumes on the environment, it is necessary to describe accurately the characteristics of the environment that could be affected by the plume before the dredging activity takes place. That is, the species and habitats present, their sensitivity (designation), tolerances and acceptable thresholds for change. The key baseline environmental characteristics of concern in this instance are water quality, marine ecology, fish and other legitimate uses of the site.

In the case of water quality, the main interest is the natural background concentration of suspended solids. This may vary from virtually zero in a coral reef area (where even a small increase due to dredging may be unacceptable) to very turbid estuaries such as the UK's Severn Estuary, where species are tolerant of turbid conditions. It is essential to be aware that background suspended solids concentrations vary greatly over time scales ranging from a few seconds to annual seasons. With particular reference to aggregate dredging, it has been noted that the turbidity generated by storms is much greater and much more widespread than that typically associated with an extraction operation.

The baseline environment can usually be described by a review of existing data and information specific to the site and, if necessary, the collection of new data via surveys. Depending on the quality and amount of existing data available, basic survey requirements might include marine benthic community surveys and fisheries surveys.

It is important to ensure that the baseline information is provided over a sufficient geographical area to cover the area of expected influence of the sediment plume (see Step 2). For example, baseline information might be required to cover the extent of tidal excursion.

Step 4 – Predicting the environmental effects

Once the baseline environment has been described, and the generation and movement of the plume predicted, an assessment of the potential environmental effects of the dredging-induced sediment plume can be made. Essentially this step comprises consideration of the impact of the predicted plume on several environmental parameters. As a first step, this should be undertaken against a detailed checklist of all potential implications (see Chapter 5) and constitutes "impact scoping". Furthermore, the assessment process should consider not just the direct and immediate effects associated with the plume, but potential cumulative impacts that might arise in combination with the effects of other projects. In many cases, a desk assessment would be expected to be sufficient; however, in sensitive cases environmental modelling may be required.

Once the effects likely to arise have been determined, it is essential to consider the extent to which they will have a significant effect; that is, will the change constitute an impact? This is likely to require quantification of the change and its comparison to background tolerances and thresholds. This detailed "impact assessment" stage, which focuses on the key issues, is more likely to involve modelling and possibly (laboratory or field) measurements. If the initial desk assessment does not reveal any likely significant effects, then this stage may be unnecessary. Again, the sensitivity of the receiving environment is paramount.

Such assessments are mostly based on environmental parameters and threshold criteria reported in the scientific literature and knowledge gained through experience of similar projects. However, the environmental effects that are reported in the literature are often based on experimental studies or projects that are not related to dredging but can give rise to similar circumstances, such as elevated suspended sediments. When using this information, it should be recognised that the effect reported might be based on an extreme situation or one that is not directly relevant to dredging. In these circumstances, care must be taken to ensure that potential effects due to dredging, and specifically the plume, are not overstated. This situation gives rise to the need for specific knowledge about the effects of dredging-induced sediment plumes, including field measurements, such as the CEFAS studies described in Section 5.5.4.

Step 5 – Determining whether the effects are acceptable

This step in the decision-making process draws on the results derived from the impact assessment undertaken in Step 4. If the answer is "yes, the change is acceptable" then some sort of approval or licence should be granted. In this case, it is good practice for the implemented project to include monitoring of the actual, as opposed to the predicted, effects. The latter is sometimes termed the "impact hypothesis". The monitoring does not need to be a comprehensive review of all parameters but should aim to test the key components of the impact hypothesis. This provides feedback to the regulating authority and to the scientists so that the predictive techniques may be refined.

If the answer is "no, the change is not acceptable" then mitigation measures may be considered. Again, this should be based on a checklist of viable and accepted mitigation approaches and will lead to a re-assessment of the acceptability of the residual effects. If they are, then acceptable permission may be given to undertake the works on a conditional basis.

7.2 GAPS IN CURRENT KNOWLEDGE

This section discuss the availability and quality of field data for future research, impact definition and model calibration, and the limits of the technology for obtaining relevant data, thereby identifying gaps in the present state of knowledge. It focuses on defining the data and tools required to inform an assessment framework.

It will be clear that consistent, validated field data are essential to both inform our knowledge of the "functioning" of the sediment plume and of the actual impacts that can arise from the changes that result. The priority for research, however, has to be the physical consequences of plume generation, because the biological effects flow from the physical change. Yet it is worth noting that a change does not represent an impact unless it has an adverse effect on a receptor species or habitat; therefore, definition of the point at which an adverse effect occurs is also important.

7.2.1 Physical processes

Consistent source term field data

As regards the modelling of plumes arising from dredging, the most significant gaps in knowledge apply to the source terms of sediment losses from dredging. The field data that have been collected hitherto have not generally been measured on a consistent or reproducible basis. The absence of reliable field data has created uncertainty in the development of dredging process models and of dynamic plume models (in the case of the trailing suction hopper dredger), and has hindered their application. Some field measurement studies have been carried out in a rigorous and self-consistent fashion, and considerable insights can be gained from these. However, there is a need for a consistent and effective measurement protocol that can be followed in the future by all researchers in this area to form a database (of published, public and private information). It could be used in the future development of models and therefore would inform decisions about acceptability. The form such a protocol might take is discussed in Section 7.2.2.

The field measurements that have been carried out on board trailing suction hopper dredgers for the aggregate dredging industry are less subject to this problem because of the different nature of the measurements and the consistency of types of operation and the dredging environments. However, the usefulness of such measurements is compromised if dynamic plume models are not sufficiently calibrated to make use of this information, and so the need for consistent sediment concentration field data still arises.

The other gaps in our knowledge of physical processes relate primarily to the processes occurring in the initial stages of plume development, namely:

- mechanical disaggregation of sediment as a result of the dredging process
- stripping of sediment from the plume during its rapid descent and the impact of the dynamic plume on the bed
- flow of the dynamic plume along the bed as a high-concentration suspension and the re-suspension of sediment from this layer into the water column
- settling velocity of the sediment in the plume at the start of the passive plume phase.

Filling these gaps is seen as a priority for research that will give the most benefit to the applied modelling of dredging plumes. Further research into other aspects of the physical processes discussed in this document will bring about improvement in the prediction of plume behaviour, but the benefits may not be so great.

Measurement of characteristics of the dynamic plume

The first of these gaps in knowledge, mechanical disaggregation of sediment by the dredging process, requires field sampling and particle sizing analysis of samples before and after dredging. Collection of such field data could form a part of a measurement protocol, discussed further in Section 8.3.1.

The second of these gaps in knowledge, the stripping of sediment from the dynamic plume, will benefit from increased rigour in field data collection. However, the most appropriate method for measuring these data, the sediment flux approach (Section 3.3.5), may not be sufficiently close to the dynamic plume to distinguish accurately between the re-suspension of sediment from the rapid descent and that from other sources. It might be necessary, therefore, to carry out specific field investigations or laboratory work to investigate dynamic plume processes. These would not need to be based on an established protocol but merely targeted towards the investigation of the dynamic phase.

The third gap in knowledge is the behaviour of high-concentration suspensions. This is being addressed by the COSINUS project, which has been researching the interaction of cohesive sediment in suspension and on the bed. The project has involved extensive fieldwork, laboratory experiments and complex mathematical modelling, as well as the development of new approaches from all of these avenues for use in applied modelling. One of the foci for this work is the behaviour of concentrated suspensions. The project is likely to do much to progress the understanding of concentrated suspensions, so until this work has been reported (due in September, 2000) and assimilated, the formal establishment of research work in this area would be ill-advised.

The settling velocity of plumes is another area that will benefit from further fieldwork, arising from both the proposed new measurement protocol and from targeted field or laboratory work.

7.2.2 A measurement protocol

Both impact assessment and the development of predictive techniques are hindered by the lack of an adequate description of the plume and the parameters causing and influencing it. This scoping study has highlighted the need for a detailed measurement protocol to permit direct comparison of plumes resulting from dredging and to allow further development of predictive models. It is not suggested that this should be applied to every dredging operation but to a sufficient number of projects to enable researchers to obtain good quality information for calibration of models.

The existence of a useable protocol demands appropriate equipment for making measurements. Because turbidity is highly variable over time and space, measuring techniques have to be able to sample at a sufficient accuracy and frequency. Measurement in both time and space to a known standard of measurement is required.

7.2.3 Environmental effects

There is a clear lack of site-specific field data on the environmental effects of dredging-induced sediment plumes in the UK. Much of the research cited in Section 5 is based on overseas work, the effects of similar though not dredging-related activities, laboratory investigation and prediction based on experience.

In addition to measuring baseline environmental data in the field, in some cases it will be important to verify the predicted environmental effects against targeted field data. Often a desk assessment or modelling can be used to predict environmental effects satisfactorily. However, the accuracy of such predictions cannot be verified without field monitoring of actual effects. This approach is likely to become more important with the recent requirement for dredging projects to consider cumulative environmental effects (see Section 5.7).

CEFAS's recent studies into the effects of sediment plumes at the Race Bank aggregate dredging site, off eastern England (see Section 5.5.4), demonstrate the potential benefit of field observations. This example is particularly relevant because the results show the extent of dredging-related effects and the influence of naturally occurring turbid conditions, and the impact of other seabed activities such as beam trawling.

Increased knowledge of actual environmental effects through field measurements should lead to an improvement of desk assessment and modelling techniques. In particular, validated field evidence should facilitate the approval process by answering questions about the acceptability of environmental effects.

It is worth restating that it is impractical to measure the response of all species and habitats to all potential change. It is suggested that field measurements focus on providing information that will allow the derivation and acceptance of generic assessment standards, based on:

- threshold concentrations for suspended sediment/turbidity and sedimentation levels
- tolerance levels of key species
- identification and response of indicator species.

Of course, innumerable thresholds and potential indicator species would need to be defined to cover the variability of the plume material, bed material, plume source and the environmental interest of concern (eg water quality parameters, marine ecology, fish and shellfish species). Based on the findings of Sections 5.2–5.7, the following list indicates the types of environmental effect for which impact predictions could most benefit from field measurements:

- plume effect on oxygen demand
- plume effect on the mobilisation of contaminants
- plume tolerances of water column communities
- plume tolerances of pelagic and demersal fish larvae and spawn
- plume effect on fish health
- plume tolerances of visually feeding fish
- plume tolerances of migrating fish
- sedimentation tolerances of seabed benthic communities
- sedimentation tolerances of estuarine and shallow littoral benthic communities
- sedimentation tolerances of shellfish, including filter-feeding species.

7.2.4 Environmental windows

Further information about the site-specific factors used to determine environmental windows is required to mitigate potential impacts through their wider application in the UK. Although there has been some detailed work defining and implementing environmental windows for UK dredging operations (see Section 6.2.2), further research is required to increase their use.

7.3 IMPROVING MODELLING TECHNIQUES

The application of the assessment framework depends to some degree (a large degree for sensitive sites) on the reliability of predictive modelling techniques. This in turn relies on the availability of good field data. To this end, and in the context of sediment plumes, models need to be refined, developed and calibrated in the areas below.

Source term (process) models

Models are required that will predict the release of sediment from different dredging plant in different operating conditions. This relies absolutely on the availability of good calibration data.

Plume dispersion models

Existing models are useful but lack calibration and development due to the absence of high-quality, consistent field data.

Cohesive sediment models

The availability of models in this area is limited (but see Section 8.3.6).

Water quality models

Again, those models that exist require parameterisation if they are to improve prediction of the levels of contamination associated with dredging and sediment plumes.

Any future guidance document should contain a review of the generic techniques and the specific models available to assist in their wider applicability and acceptance. This has been partly done in this scoping study (see Appendices 5 and 6) but should be extended in a further phase.

8 Recommendations – the way forward

8.1 DEVELOPING A COMPREHENSIVE ASSESSMENT FRAMEWORK

CIRIA Research Project 600 was established to review the current state of knowledge on the nature of sediment plumes arising during dredging operations and their effects on the environment. In addition, a key objective of the project was to recommend further work required to:

- produce a comprehensive framework for assessing the effects of sediment plumes arising from dredging
- address knowledge gaps, including modelling limitations
- produce best practice guidance on the assessment of effects.

In accordance with this objective, this chapter sets out recommendations for progressing the science of assessing the effect of dredging-induced sediment plumes in the future.

The recommendations set out below aim to address the gaps in current knowledge identified in Section 7.2 and to provide the future research necessary to achieve a comprehensive, widely accepted assessment framework and better knowledge transfer. There are three steps towards achieving this.

1. Adopt an assessment framework that indicates the method steps required to make an assessment and the information required to inform the decisions (Section 8.2).

2. Fill the knowledge and technology gaps in the assessment process through research, development and monitoring (Section 8.3).

3. Produce good practice guidelines on the assessment of sediment plumes, the structure of which should be based on the draft framework, and the outcome of which will be a comprehensive assessment system based on validated field data (Section 8.4).

In line with these needs, Figure 7.1 provides an indicative assessment framework. In this context, the main requirements to inform the assessment of sediment plumes arising from dredging are model calibration data and the quantification of actual environmental effects. However, it is unrealistic to expect every dredging operation to undertake detailed plume and environmental effects monitoring, so detailed generic studies are required to inform the models and predictions of environmental effects. The agreement and adoption of monitoring protocols would significantly improve the use and effectiveness of monitoring data collected in the future. Absolute standards, however, should not be proposed. These ideas are developed further in the following sections.

8.2 ADOPTION OF A DRAFT ASSESSMENT FRAMEWORK: RESEARCH PRIORITIES

It is recommended that the indicative assessment framework presented in Section 7.1 be adopted. It presents a logical approach to determining the turbidity a particular dredging project will cause and the acceptability of the impacts of that turbidity. The framework provides the basis for identifying gaps in knowledge and prioritising future research.

The first research priority in this context is the ability to describe the turbidity plume comprehensively. This means improving both the quality of the baseline data and the modelling techniques, as described in Sections 7.2 and 7.3 respectively.

It is not economically practical to measure continuously the turbidity generated during every dredging operation. It is essential, therefore, to develop **predictive techniques** for stakeholders to use in determining the likely acceptability of a particular operation.

Although there is no great difficulty in predicting the type of plume (dynamic or passive) that will arise from a particular dredging operation, it is not yet possible to provide accurate quantitative assessments. The immediate problem is in making an accurate estimate of the quantity of sediment released into the water during dredging – what we have called "the source term". Most, if not all, previous measurements have been made some distance away from the dredger using measuring techniques of variable reliability. Calibration of the equipment is also suspect in many cases. This creates two major problems: first, in most cases, no one knows what the actual turbidity is; second, there is therefore no reliable calibration data available from which confidence in modelling techniques could be improved.

8.3 ADDRESSING KNOWLEDGE GAPS

8.3.1 Development of an internationally accepted monitoring protocol

There is a pressing need to develop an internationally accepted protocol for taking measurements of suspended sediment concentrations around a dredger. The protocol should be developed in consultation with UK and international stakeholders, including those described in Appendix 2. It will need to cover all types of dredging operation and plant normally likely to be encountered. The form that such a protocol might take is addressed below.

As part of Phase II of the VBKO research initiative (see Section 2.2), suggestions were advanced for a protocol for collecting data so as to develop dredging process models. The document concluded that the steady state approach, exemplified by the S-factor approach, is impractical because it is difficult to measure concentration distribution of the entire plume and because the present understanding of the settling velocities in plumes is uncertain. The document considered that the best way forward was the sediment flux approach, which involves the simultaneous measurement of velocity and suspended sediment concentration across the plume cross-section.

The Øresund Link measurements provides an example of this approach and probably offers the most comprehensive set of measurements to date. The sediment flux method represents the most practical method of enabling different dredging operations in varying conditions to be compared directly. However, there are some problems with this method that will need to be addressed before it forms the basis of a viable protocol.

One such problem is the choice of techniques for measurement. The method requires the simultaneous measurement of velocity and concentration profiles across the plume. However, conventional current and silt meters can only make point measurements (and cannot be used rapidly enough to make more than a few measurements) through the vertical and this may not be enough to resolve the sediment flux sufficiently. The alternative is to use acoustic profilers. These can measure the simultaneous distribution of velocity and concentration instantaneously through the water column. The rapid profiling that these methods allow is particularly suitable for profiling the cross-section of a plume. However, there are some drawbacks:

- the acoustic method does not provide data in the upper 1–1.5 m and the bottom 6–15 per cent of the water column; the bottom of the water column is particularly important, as this is where the greatest concentrations of sediment tend to occur
- acoustic techniques are subject to limitations, particularly those arising from air bubbles entrained in sediment plumes, which may preclude or limit their use in some situations.

The technology exists to overcome these problems. However, the complicated nature of the technology means that only a few institutions are likely to know how to operate the systems successfully. A way must be found to progress this issue without compromising the quality, and therefore usefulness, of the measurements.

8.3.2 Field-testing suspended sediment monitors

Field tests should be carried out to investigate and demonstrate the viability of using newer techniques alongside more conventional techniques to measure suspended sediment concentration adequately.

8.3.3 Determine a standard procedure for measuring the disaggregation index

Key to understanding the generation of turbidity, is predicting the response of various types of sediment to a variety of dredging techniques. No standard soil mechanics test exists that can indicate the extent to which sediment will disaggregate under mechanical and hydraulic forces. Work is required to develop a procedure for testing sediment to obtain the relevant properties. This has been termed "the disaggregation index".

It is recommended that laboratory tests be carried out to develop a standard procedure for determining the disaggregation properties of sediment to be dredged.

8.3.4 Monitoring benthic boundary layer processes

Although acoustic profilers are widely used to monitor sediment plumes, they miss crucial parts of the water column, namely the 6–15 per cent nearest the seabed and the surface layers. These areas are likely to be the most important in terms of biological activity and therefore those where biological impacts are most likely to occur. In these benthic boundary layers other processes (such as the effects of waves, increased turbulence due to seabed topography and gravity currents, and turbulence dampening due to deposition) may affect the suspension of the sediment plume. These process can increase the area of the sediment plume footprint, but the magnitude of their impact on sediment plumes is poorly understood. It is, therefore, recommended that monitoring is undertaken to improve understanding of the impact of bathymetry on sediment plume dynamics. One possible approach to this is the use of minipods (see Section 5.5.4).

8.3.5 Monitoring environmental effects

As identified in Section 7.2.3, there is a general lack of field data on the environmental effects of dredging-induced sediment plumes. Targeted field surveys are needed to determine whether (predicted) impacts arising from sediment plumes are accurate and to quantify their significance. Of particular interest are measurements relating to turbidity, plume longevity and sedimentation patterns, species response and habitat response.

In the UK, the monitoring of the environmental effects of sediment plumes has only recently been undertaken in detail, through the deployment by CEFAS of minipods in Area 107 of the Race Bank aggregate dredging site (see Section 5.5.4). Consideration of the minipod data demonstrates that field monitoring improves our understanding of the scale of environmental effects. With reference to Box 5.4, the field data from the minipods provided guidance on the extent of potential environmental effects.

It is recommended that further field monitoring is undertaken to improve understanding of the environmental effects of sediment plumes, particularly with respect to species and habitat response, and therefore to improve the predictions associated with environmental impact assessments.

8.3.6 Database

There should be an investigation into the practicability and cost of compiling a database of all the available measuring data on the local flow, wave and turbidity regime. This database should at least indicate the type and amount of data available as well as the source and location of these data. It would allow a quick first assessment of the plume dispersion conditions to be made.

8.3.7 Filling gaps in modelling capability

Recommendations for the development and refinement of relevant models are discussed below. The next stage will be to use the field and laboratory data obtained to refine existing models. New models to predict source terms are already in existence but lack calibration data. Significantly larger gains can now be made through calibrating the models available based on good data, rather than refining the models themselves. The data obtained will provide a means of determining coefficients in the algorithms used in those models and will perhaps highlight the need to include other parameters.

Source term (process) modelling

Virtually all plume models require a source term to be specified. Models are required that will predict the amount and rate of release of sediment based on the type of dredging plant, the operating conditions and the material being dredged. A few such models exist, mainly in the USA, and steps towards this have been taken in the VBKO project. The usefulness of such models depends entirely on obtaining good calibration data. Until these data are obtained, the validity of the modelling approach cannot be fully tested and the plume models themselves can give only relative, not absolute, numerical predictions.

Plume dispersion model calibration

The scoping study has reviewed the available models and found that several of these can be and are used for applied modelling of plume advection and dispersion. However, the models suffer from a lack of input data (see Section 7.3 above), lack of calibration and lack of development due to the lack of good quality field data. For these models, the greatest benefit will come from the formulation of a measurement protocol, as described in Section 7.2.2, and the increased availability of such data.

Cohesive sediment modelling

Cohesive sediment modelling represents an area where the development of models would be beneficial, in particular, for those concerned with the behaviour of concentrated suspensions and the related formation of fluid mud. This is the subject of a major European research project (COSINUS). It is expected that the results of this project will provide a major improvement to modelling in this area and to cohesive sediment modelling in general.

Water quality models

Models also exist that can be and are used for the prediction of water quality impacts. However, these models do not reproduce the dynamic interaction of contaminants with water and sediment, instead taking a simpler approach to processing dispersion model results. By far the greatest uncertainty in the models' results is associated with the selection of key contaminant parameters from the literature to describe what is an extremely site-specific process. The choice of parameters can affect predicted concentrations of contaminants by orders of magnitude greatly exceeding any inaccuracy due to the modelling approach. The development of water quality modelling in the context of dredging plumes should focus on more closely identifying the contaminant parameters that apply to the site-specific conditions of dredging.

Model robustness

Care must be taken to ensure that the models used to predict sediment plume dynamics are robust. Models should be calibrated or validated with observations from the dredge site. Clear guidance should be given on whether the models in use are site-specific or generic. Modelling must also be appropriate to the hydrodynamic and bathymetric environment in question. Open-sea data, for example, is often only as accurate as the sensitivity of many models; similarly, the grid scale used to represent complex bathymetry should not be too coarse. Bathymetry must be well represented in plume prediction models – an aspect that has received little attention in the past.

In summary, it is recommended that, as a priority, calibration data are obtained from field measurements and used to upgrade existing computer models of source terms. Furthermore, the collection of these data must be based on a recognised, adopted measurement protocol.

8.3.8 Use of environmental windows

As identified in Sections 6.2 and 7.2.4, the use of environmental windows to control and limit the timing of UK dredging operations has been limited to date, particularly with respect to the direct mitigation of environmental effects arising from sediment plumes.

It is recommended that information on the use of environmental windows in the UK is collated and knowledge of its effectiveness is disseminated to increase awareness. A manual on best practice could include detailed examples of previous and current uses of environmental windows. The examples should demonstrate how the windows were identified, and what the determining factors were. They may need to be specific to aggregate and capital and maintenance dredging.

8.4 DEVELOPING BEST PRACTICE GUIDANCE FOR ASSESSING ENVIRONMENTAL EFFECTS

It is recommended that best practice guidance be produced on the assessment of the environmental effects of sediment plumes arising from dredging. This document should provide sufficient information to inform the decision steps detailed in Figure 7.1, by addressing, for example, the conditions under which a plume will arise and the likely nature of that plume in different circumstances. There should be information on dredging techniques, sediment types, hydrodynamic conditions and the baseline environment. The guidance should also provide details of the environmental effects that arise in different circumstances, the factors determining the point at which the effect represents an impact, and practical means to reduce the impact. These issues are expanded upon in the following sections.

8.4.1 A guidance context

Predicting the environmental effects of sediment plumes is a complex task, particularly since it relies on accurate predictions of plume processes. Furthermore, predicting the potential physical, biological and chemical effects of sediment plumes involves consideration of the interactions between numerous environmental components, many of which are not well understood. The problem is exacerbated by the difficulties experienced in making direct observations of cause and effect relationships in the field.

An understanding of sediment plumes facilitates the assessment of environmental effects, the determination of mitigation measures and the definition of monitoring requirements. At present, knowledge is based on uncoordinated field measurements, laboratory tests and theoretical models. Because little is known about site-specific environmental effects, they can be predicted only in generic terms and the real extent of an effect often remains unknown. Again, there is a need for consistent field measurements of plume dynamics, turbidity, sedimentation patterns, and species and habitat response.

The improved prediction of environmental effects will benefit the dredging industry's stakeholders. For example, it has been recognised that the UK aggregate dredging industry would benefit from environmental impact assessments supported by accurate site-specific investigations, particularly if this would improve the precision of monitoring requirements relating to licence conditions. Also, the potential impacts of sediment plumes on the migration of anadromous fish is frequently cited as being of concern in the USA, yet there is a lack of quantitative evidence for such an impact in peer-reviewed scientific literature. Measured evidence of environmental effects, therefore, will help to reduce reliance on a precautionary approach.

With the above in mind, this section considers several aspects of predicting sediment plumes and associated environmental effects that would benefit from widely adopted best practice. It is recommended that best practice guidance should be based on the assessment framework described in Section 7.1, such that an assessment follows the decision-steps shown in Figure 7.1 and the guidance provides the information and tools to enable well-informed decisions and predictions to be made.

8.4.2 Understanding environmental effects

To develop best practice, it is clear that better guidance is required on understanding and predicting likely environmental effects. Ideas regarding the content of future guidance are described below. The issues identified are based on data gaps and assessment limitations identified throughout Sections 5.2 to 5.6 and in Section 7.2.3.

Guidance on predicting environmental effects

In order to achieve accurate predictions of environmental effects, quantitative field evidence should be obtained (eg the effect of a plume on fish migration). Field research is needed to assess whether dredging-induced sediment plumes cause significant effects compared to natural sources of suspended sediment, such as storms.

Where possible, the assessment of a dredging operation should be supported by **site-specific** field monitoring data. So far, field studies of the environmental effects of plumes have been carried out at only a few dredging sites. At other sites, just the likely effects have been predicted, but, given the present state of field data and the commitment involved in monitoring, the latter approach is necessary. It would be inappropriate to require every dredging operation to undertake detailed plume and habitat response monitoring. Research should, however, include field measurements of actual environmental effects at a greater range of sites. Furthermore, this approach needs to be incorporated into a co-ordinated field monitoring programme.

The development of targeted monitoring objectives are essential to develop a field monitoring programme focused on those environmental parameters that are most likely to be affected by dredging at the site in question. Information resulting from monitoring should then be made publicly available and fed into future environmental impact assessments, where appropriate. This approach should increase the accuracy of predicting environmental effects.

It is not appropriate to monitor for environmental effects that are unlikely to occur or unlikely to be significant. Nor is it practical to monitor the effect of every aspect of the dredging operation on every environmental parameter. For water quality, field monitoring should focus on physical, chemical and possibly biological thresholds. For biological parameters, field monitoring should focus on indicator species, particularly those of commercial interest. As it is not feasible to determine actual effects on all aspects of the environment, generic tolerance thresholds and indicator species should be carefully identified and selected. In some cases, biological indicators will be site-specific, such as the use of eelgrass for the Øresund Link.

Guidance on field monitoring

Monitoring should be undertaken to test the predictions made in environmental impact assessment and cumulative effect assessments. Monitoring programmes should be designed to form an active part of the management of the dredging operation. Sufficient information should be obtained initially on baseline environmental conditions, to facilitate the setting of thresholds or limits of acceptable change (ie tolerances).

In this context, a relevant threshold might be a restriction on suspended sediment concentrations, such that they must not increase by more than a certain percentage above normal background concentrations during the dredging operation (taking account of natural variation). The results of the monitoring should then be fed to the operational managers. This will allow the managers to put mitigation measures in place to help prevent the thresholds of acceptable change being exceeded and therefore avoid any adverse environmental impact. Monitoring must also take into account any natural variation (eg seasonal conditions) and one-off events (eg storms).

8.4.3 Guidance on the assessment of sediment plumes: impact prediction

Impact assessments considering the effect of sediment plumes arising from dredging would benefit from the availability of detailed information on potential environmental effects; that is, impact checklists. Such checklists should combine current knowledge on the characteristics of different environments, the various dredging techniques and the behaviour of different sediment types, and would have to acknowledge these differences. For example, different environmental effects will obviously arise from maintenance dredging in cohesive silt in a turbid estuarine environment, compared with aggregate dredging for gravel in clear open seas.

Checklists cannot cover site-specific environmental effects, but they can be effective in enabling the user to predict whether the sediment plume might affect environmental interests. In the UK, checklists might best be applied to aggregate dredging since many dredging areas are similar. A checklist would need to be supported by a guidance document, which should cover the various environmental scenarios that can influence the environmental impact. Guidance is required on factors such as:

- indicator species
- suspended sediment tolerant/intolerant species
- impact significance
- cumulative environmental effects.

So far as is practicable, the guidance should bring together the latest information on environmental effects and inform the user about field research and experience.

It is recommended, therefore, that best practice guidance on the assessment of the effect of sediment plumes be produced in the near future. The current state of knowledge is sufficient to provide good advice on the approach that should be adopted and the data and tools available. The adoption of such guidance would meet the urgent need for standardised, accepted methodologies for data collection, monitoring and impact prediction. It is essential that all relevant stakeholders participate in the production of this guidance. Once produced, the guidance (incorporating details on likely impacts) should be regularly updated, disseminated widely and subject to continuing professional development in light of further research to fill knowledge/data gaps. This process would provide an important focus for feeding back the results of monitoring.

8.4.4 Development of an indicative environmental effects framework

The indicative framework provided in Section 7.1 should be developed to improve prediction of environmental effects. Such development should include compilation of comprehensive lists of the potential individual and cumulative environmental impact (on water quality, marine ecology and fisheries and shellfisheries) of the dredging-induced sediment plume. Good information and thorough research should support these lists, which should acknowledge the variety of effects that arise in different circumstances.

Similarly, specific details of potential mitigation measures are also needed. For example, it is important to include information about the suitability of adapting dredging plant to reduce sediment re-suspension and the suitability of environmental windows for protecting particular species.

In the short term, the development of an indicative framework should focus on detailing those environmental effects and mitigation measures associated with sediment plumes from a single operation. In the longer term, the work must be developed to provide details of typical cumulative impacts. The consideration of cumulative effects may, in fact, require the development of a separate framework. There is a clear need to develop a consistent approach to considering the cumulative effects of works adjacent to or within European conservation sites (ie SACs and SPAs). Such an approach could take the form of a dredging industry protocol, and would have to take account of past, present and future impacts as well as the influence of other works. Such an initiative can only progress, however, if the limits of responsibility are set and responsibility for information provision is allocated.

8.5 RECOMMENDATIONS FOR KNOWLEDGE TRANSFER

8.5.1 Protocol publicity

Once an assessment framework and measuring protocol have been accepted, the international dredging industry should be persuaded to monitor dredging operations and make the information available to researchers developing models and predicting environmental effects.

Efficient and effective knowledge transfer will be an essential part of publicising the need for measurements to be made based on agreed protocols and techniques. It will also be important in seeking co-operation in taking measurements and making the data publicly available.

Knowing re-suspended sediment concentrations, the nature of the sediment plume likely to be generated by a particular dredging operation and how it is transported form only the first stage in the assessment. It is recommended that the second priority is the need to improve knowledge of the actual environmental effects of re-suspended sediment – relating the cause to the effect. It is recognised that this may be very site-specific, so the emphasis should be on determining correlations using targeted field measurements to support laboratory techniques.

8.5.2 Environmental effects checklist and guidance

The value of environmental checklists and associated guidance will depend on knowledge transfer. Greater use will provide more incentive and information for updating, and a mechanism for dissemination. The wider availability of new knowledge, best practice and new technologies will in turn facilitate the continual improvement of the standard of environmental impact assessment and cumulative effect prediction.

References

ADEY, W H and McKIBBIN, D L (1970)
Studies on the maerl species Phymatolithon calcareum (Pallas) nov. com. and Lithothamnium coralloides (Crouan) in the Ria de Vigo
Botanica mar, vol 13, pp 100–106

ANDERSIN, A-B and SANDLER, H (1983)
Recent changes in the macrozoobenthos communities in the deep sea areas of the Baltic proper and the Gulf of Finland
Paper held at the 8th Symposium of the Baltic Marine Biologists, Lund, 10–14 Aug. 15 pp

ANDRASSY, C and HERBICH, J B (1988)
Generation of re-suspended sediment at the cutterhead
Dock and Harbour Authority, vol 67, no 797, pp 207–216

BARNARD, W D (1978)
Prediction and control of dredged material dispersion around dredging and open water pipeline disposal operations
Technical Report DS-78-13. US Army Engineer Waterways Experiment Station, CE Vicksburg, Mississippi. 112 pp

BARNES, R S K and HUGHES, R N (1988)
An Introduction to Marine Ecology
Second Edition, Blackwell Science

BATES, A (1996)
Reopening the navigation channel to the Port of Londonderry
In: *Proceedings of the Institute of Civil Engineers, Water Maritime and Energy*, vol 113, no 1

BLOKLAND, T (1986)
The turbidity caused by various dredging techniques: an analysis of measurements in Rotterdam and Delfzijl (in Dutch)
MKO 110.18-R8647, Gemeenteweken Rotterdam Ingenieursbureau Havenwerken, the Netherlands

BLOKLAND, T et al (1986)
Environment conscious dredging and disposal, turbidity measurements Delfzijl (in Dutch)
Co.110.18-R8631, Minimalisering Kosten Onderhoudsbaggerwerk – Dredging Research Association

BLOKLAND, T (1988)
Determination of dredging-induced turbidity
Terra et Aqua, vol 38, pp 3–12, International Association of Dredging Companies (IADC), the Netherlands

BMT (1996)
Nash Bank Environmental Assessment for aggregate extraction from Nash Bank, Bristol Channel
Prepared for United Marine Dredging Ltd, ARC Marine Limited and British Dredging Aggregates Ltd, May 1996

BOSLAND, P (1991)
Troebelheidsumetiugen rond Wormwielzuiger in de Geulhaven te Rotterdam
Report J644 to Van Oord-Werkendam BV

BRAHME, S B and HERBICH, J B (1986)
Hydraulic model studies for suction cutterheads
ASCE Journal of Waterway, Port, Coastal and Ocean Engineering, vol 112, no 5, September 1986

BRAY, R N, BATES, A D, and LAND, J M (1997)
Dredging. A handbook for engineers
Second edition. Arnold, London

BROWN C L and CLARK R (1968)
Observations on dredging and dissolved oxygen in a tidal waterway
Wat Resour Res, vol 4, pp 1381–1384

BRYAN, G W, GIBBS, P E, HUMMERSTONE, L G and BURT, G R (1986)
The decline of the gastropod *Nucella lapillus* around south-west England: evidence for the effect of tributyltin from antifouling paints
J Mar Biol Ass UK, vol 66, pp 611–640

CANADIAN ENVIRONMENTAL ASSESSMENT AGENCY (1999)
Cumulative Effects Assessment Practitioners Guide
Prepared by the Cumulative Effects Assessment Working Group and AXYS Consulting Ltd. February, 1999

CARSLAW, H S and JAEGER, J C (1959)
Conduction of Heat in Solids
Second edition. Oxford University Press (Clarendon), London and New York

C-CORE (1996)
Proposed marine mining technologies and mitigation techniques: a detailed analysis with respect to the mining of specific offshore mineral commodities
Contract report for US Department of the Interior, Minerals Management Service. C-CORE publication 96-C15

CEFAS (1998)
Impact of dredger plumes on Race Bank and surrounding areas – Short Version.
Internal CEFAS report, 1998

CEQ (1997)
Considering Cumulative effects under the National Environmental Policy Act
Council on Environmental Quality, January 1997

COLLINS, A (1995)
Dredging-induced near-field re-suspended sediment concentrations and source strengths
USACE WES Miscellaneous Paper D-95-2, August

De GROOT, S J (1986)
Marine Sand and Gravel extraction in the North Atlantic and its Potential Environmental Impact, with Emphasis on the North Sea
Ocean Management, vol 10 (1988) pp 21–36

DOER (1999)
Estimating dredging sediment resuspension sources
DOER – E6

DONZE, M (ed) (1990)
Aquatic pollution and dredging in the European Community
Eleventh Lustrum of the Association of Dutch Dredging Contractors, The Hague. DELWEL

DOTS (1986)
Guide to selecting a dredge for minimising resuspension of sediment
Report EEDP-09-1

DOTS (1988)
Sediment resuspension by selected dredges
Report EEDP-09-2

DREDGING AND PORT CONSTRUCTION (1999)
International Directory of Dredgers

DYER, K R, CORNELISSE, J, DEARNALEY, M P, FENNESSY, M J, JONES, S E, KAPPENBURG, J, McCAVE, I N, PEJRUP, M, PULS, W and WOLFSTEIN, K (1996)
A comparison of in-situ techniques for estuarine floc settling velocity measurements, Netherland
Journal of Sea Research, vol 36, part 1, pp 15–29

EMU ENVIRONMENTAL (1998)
Inner Owers Environmental Assessment for aggregate extraction from Inner Owers, English Channel
Prepared for United Marine Dredging Ltd and ARC Marine Limited, January 1998

ENVIRONMENT CANADA (1994)
Environmental impacts of dredging and sediment disposal
By Les Consultants Jacque Bérubé Inc for the Technology Development Section, Environmental Protection Branch, Environment Canada, Quebec and Ontario Regions. Cat. no En 53-39/1994E. 97 pp

ERM (1996)
Environmental Impact Assessment Study for Disposal of Contaminated Mud in the East Sha Chau Marine Borrow Pit, Initial Assessment Report
ERM-Hong Kong Ltd Report, July

ERM (1999)
Aggregate Production Licence, Middle Bank, Firth of Forth: Environmental Statement
Prepared by Environmental Resources Management for Westminster Gravels Ltd

EVANS, C D R, CROSBY, A, WINGFIELD, R T R, JAMES, J W C, SLATER, M P and NEWSHAM, R (1998)
Inshore seabed characterisation of selected sector of the English coast
Technical Report WB/98/45. Keyworth: British Geological Survey

EVANS, S M, LEKSONO, T and McKINNELL, P D (1995)
Tributyltin pollution: a diminishing problem following legislation limiting the use of TBT-based anti-fouling paints
Marine Pollution Bulletin, vol 30, no 1, pp 14–21

EVERHART, W H and DUCHROW, R M (1970)
Effects of suspended sediment on aquatic environments
NTIS US Department of Commerce, PB-196-641

FEATES, N G, HALL, J R, MITCHENER, H J and ROBERTS, W (1999)
COSINUS Field Experiment, Tamar estuary, Measurement of properties of suspended sediment flocs and bed properties
HR Wallingford Report TR 82

GIBB (1997)
Environmental Statement for Proposed Dredging on the Helwick Bank
Prepared by Gibb Wales for Llanelli Sand Dredging Ltd

GODCHARLES, M F (1971)
A study of the effects of a commercial hydraulic clam dredge on benthic communities in estuarine areas
Florida Department of Natural Resources, Marine Research Laboratory. Technical Series no 64

GRASSLE, J F and GRASSLE, J P (1974)
Opportunistic life histories and genetic systems in marine benthic polychaetes
Journal of Marine Research, vol 32, pp 253–84

GRAVELLI, D (1995)
Application of a model for the dispersion of material re-suspended by dredging activity in the River Tees
Final training report for Techware, April to July 1995

GRIGG, D I (1970)
Some effects of dredging on water quality and coral reef ecology
Caribbean Conservation Environmental Newsletter, vol 1, pp 22–27

GRUET, Y (1986)
Spatio-temporal changes of sabellarian reefs built by the sedentary polychaete *Sabellaria alveolata* (Linne)
PSZNI Marine Ecology, vol 7, pp 303–19

HITCHCOCK, D R and DEARNALEY, M P (1995)
Investigation of benthic and surface plumes associated with marine aggregate production of the United Kingdom
Presented at US Minerals Management Service transfer meeting, December 1995

HITCHCOCK, D R and DRUCKER, B R (1996)
Investigation of benthic and surface plumes associated with marine aggregates mining in the United Kingdom
In: *The global ocean – towards operational oceanography. Proceedings of Conference on Oceanology International*
Spearhead Publications, Surrey, Conference Proceedings, vol 2, pp 1–8

HOWELL, B R and SHELTON, R G J (1970)
The effect of china clay on the bottom fauna of St Austell and Mevagissey Bays
J Mar Biol Assoc UK, vol 50, pp 593–607

HR WALLINGFORD (1989)
Grangemouth Mud Properties
HR Wallingford Report SR 197

HR WALLINGFORD (1990)
Water injection dredging of mud – field monitoring in Harwich Harbour, Phase 1
Report EX 2159 (Confidential)

HR WALLINGFORD (1993a)
Londonderry Channel Deepening. Monitoring of suspended sediments – Lough Foyle
Report EX 2813 (Confidential)

HR WALLINGFORD (1993b)
Shingle Bank Hastings. Dispersion of dredged material
Report EX 2811

HR WALLINGFORD (1994a)
Shoreham Outfall. Turbidity monitoring during dredging
Report EX 3044 (Confidential)

HR WALLINGFORD (1994b)
Great Yarmouth. Dispersion of dredged material
Report EX 2967

HR WALLINGFORD (1994c)
South Humber Bank Power Station Cooling Water Intake. Sedimentation modelling
Report EX 3058

HR WALLINGFORD (1995a)
Estimation of sediment fluxes from ADCP backscatter data
Unpublished paper PHR 118

HR WALLINGFORD (1995b)
Initial dispersion phase of plumes of sediment arising from sand dredging overflow
Unpublished paper PHR 119

HR WALLINGFORD (1996a)
Re-suspension of bed material by dredging. An intercomparison of different dredging techniques, Vol 1
Main Report, SR 461

HR WALLINGFORD (1996b)
Dispersion studies at Owers Bank
Report EX 3335 (Confidential)

HR WALLINGFORD (1997)
KCRC West Rail Project, Tidal flow and sediment plume modelling
Report EX 3683

HR WALLINGFORD and POSFORD DUVIVIER ENVIRONMENT (1998)
Harwich Haven Approach Channel Deepening. Environmental Statement
Prepared for Harwich Haven Authority, January, 1998

HR WALLINGFORD (1999a)
Environmental Aspects of Aggregate Dredging, Refined source terms for plume dispersion studies
Report SR 548

HR WALLINGFORD (1999b)
Model development for the assessment of turbidity caused by dredging: Volume 1
Technical Report EX 3998

HR WALLINGFORD (1999c)
Dynamics of Estuarine Muds, A manual for practical applications
Report SR 527

HR WALLINGFORD (1999d)
Model development for the assessment of turbidity caused by dredging, Phase 2: Part 1, Turbidity Measurement Protocol
Proposal PR 1006

IADC/CEDA (1997)
Environmental aspects of dredging. Guide 3. Investigation, interpretation and impact
IADC/CEDA. p 67

ICES (1975)
Report of the Working Group on Effects on Fisheries of Marine Sand and Gravel Extraction
ICES. Cooperative Research Report no 46

ICES (1977)
Second Report of the ICES Working Group on Effects on Fisheries of Marine Sand and Gravel Extraction
ICES. Cooperative Research Report no 64

ICES (1992)
Effects of Extraction of Marine Sediments on Fisheries
ICES. Cooperative Research Report no 182

JENSEN, A, VALEUR, J, THORKILSEN, M and WEIERGANG, J (1995)
The use of numerical modelling in monitoring of dredging spoils
In: *Proceedings of the 14th World Dredging Congress.* Amsterdam. November, 1995

JENSEN, E F P (1999)
Introduction to spill, spill monitoring and spill management
In: *Proceedings of the Øresund Link Dredging and Reclamation Conference*
Copenhagen. 26–28 May 1999

JOHNSON, B H and FONG, M T (1994)
Development and verification of numerical models for predicting the initial fate of dredged material disposed in open water – Report 2, Model development and verification
Technical Report DRP-94-x, USACE, WES, Vicksburg, USA

JOHNSON, B H, McCOMAS, D N, McVAN, D C and TRAWLE, M J (1993)
Development and verification of models for predicting the fate of dredged material disposed in open water
Draft Technical Report USACE, WES, Vicksburg, USA

JOHNSTON, S A (1981)
Estuarine dredge and Fill Activities: A Review of Impacts
Environmental Management, vol 5, no 5, pp 427–440

KAPLAN, E H, WELKER, J R, KRAUS, M G and McCOURT, S (1975)
Some factors affecting the colonisation of a dredged channel
Marine Biology, vol 32, pp 193–204

KIRBY, R and LAND, J M (1991)
The impact of dredging – A comparison of natural and man-made disturbances to cohesive sediment regimes
In: *Proceedings of CEDA Dredging Days.* Amsterdam, November 1991

KOH, R C and CHANG, Y C (1973)
Mathematical model for barged ocean disposal of waste
Technical series EPA 660/2-73/029, US Environmental Protection Agency, Washington, DC

KRONE, R B (1962)
Flume studies of the transport of sediment in estuarial shoaling processes
Final Report, Hydraulic Engineering and Sanitary Engineering Research Laboratory, University of California, Berkely, USA

LAND J M and BRAY R N (1998)
Acoustic measurement of suspended solids for monitoring of dredging and dredged material disposal
WODCON, Las Vegas

LAND, J M, KIRBY, R and MASSEY, J B (1994)
Recent innovations in the combined use of acoustic doppler current profilers and profiling siltmeters for suspended solids monitoring
In: *Proceedings of 4th Nearshore and Estuarine Cohesive Sediment Transport Conference,* Wallingford, UK

LONG, B G, DENNIS, D M, SKEWES, T D and POINER, I R (1996)
Detecting an environmental impact of dredging on seagrass beds with a BACIR sampling design
Aquatic Botany, vol 53(3), pp 235–243

LOOSANOFF, V L (1961)
Effects of turbidity on some larval and adult bivalves
Proc. Gulf Carib. Fish. Inst., vol 14, pp 80–95

LORENZ, R (1999)
Spill from dredging activities
Proceedings of the Øresund Link Dredging and Reclamation Conference.
Copenhagen 26–28 May 1999, pp 309–324

LUNZ, G R Jr (1938)
Oyster culture with reference to dredging operations in South Carolina
Report to US Army Corps of Engineers Office, Charleston, SC

McCALL, P L (1976)
Community patterns and adaptive strategies of the infaunal benthos of Long Island Sound
Journal of Marine Research, vol 35, pp 221–266

McLUSKY, D S (1989)
The Estuarine Ecosystem
Second edition, Chapman and Hall, p 215

MAURER, D L, LEATHEM, W, KINNER, P and TINSMAN, J (1979)
Seasonal fluctuations in coastal benthic invertebrate assemblages
Estuarine and Coastal Marine Science, vol 8, pp 181–193

MAURER, D L, KECK, R, TINSMAN, J, LEATHEM, W, WETHE, C, LORD, C and CHURCH, T (1986)
Vertical migration and mortality of marine benthos in dredged material: a synthesis
Internationale Revue ges. Hydrobiologie, vol 71, pp 49–63

MMS (1993)
Synthesis and analysis of existing information regarding environmental effects of marine mining
Prepared for the US Department of the Interior Minerals Management Service Office of International Activities and Marine Minerals (INTERMAR). OCS Study MMS 93-0006

MMS (1996)
Marine Mining Technologies and Mitigation Techniques
US Department of the Interior Minerals Management Service. July, 1996

MMS (1998)
Marine Aggregate Mining Benthic and Surface Plume Study
Report prepared for United States Minerals Management Service,
Report No 98-555-03, October 1998

MORTON, B R, TAYLOR, G and TURNER, J S (1956)
Turbulent gravitational convection from maintained and instantaneous sources
Proceedings of the Royal Society, Vol A234, pp 1–23

MOTT McDONALD (1991)
Contaminated spoil management study
Final Report to Environmental Protection Department, Hong Kong, under agreement CE 30/90

NEWELL, C R, SHUMWAY, S E, CUCCI, T L and SELVIN, R (1989)
The effects of natural seston particle size and type on feeding rates, feeding selectivity and food resource availability for the mussel, *Mytilus edulis* Linnaeus, 1758 at bottom culture sites in Maine
Journal of Shellfish Research, vol 8, pp 187–196

NEWELL, RC, SEIDERER, LJ and HITCHCOCK, DR (1998)
The impact of dredging works in coastal waters: A review of the sensitivity to disturbance and subsequent recovery of biological resources on the seabed
Oceanography and Marine Biology: an Annual Review, vol 36, pp 127–178

NRA (1995)
Contaminants entering the sea. A report on contaminant loads entering the seas around England and Wales 1990–1993
Water Quality Series No 24, HMSO

OAKWOOD ENVIRONMENTAL (1993)
Bristol Channel Outer Dredging Licence Application
Draft Environmental Statement. Prepared for Civil and Marine Ltd, June 1993

OAKWOOD ENVIRONMENTAL (1997)
Application for Marine Aggregate Extraction Licence: Area 432 West Varne
Environmental Statement and Non-technical Summary – Draft for Consultation
Prepared for South Coast Shipping Co Ltd, April 1997

OAKWOOD ENVIRONMENTAL (1999)
Strategic cumulative effects of marine aggregates dredging
Prepared for US Department of the Interior, Minerals Management Service

OFFICER, C B, BIGGS, R B, TAFT, J L, CRONIN, L E, TYLER, M A and BOYNTON, W R (1984)
Chesapeake Bay Anoxia: Origin, Development and Significance
Science, vol 223, pp 22–27

ONUF, C P (1994)
Seagrasses, dredging and light in Laguna Madre, Texas, USA
Estuarine, Coastal and Shelf Science, 39, pp 75–91

ØRESUNDKONSORTIET (1997)
The environment and fixed link across Øresund
Øresundkonsortiet, September 1997

ØRESUNDKONSORTIET (1998)
The Øresund Link – Assessment of the impacts on the marine environment of the Øresund Link. Update
Øresundkonsortiet, March 1998

PALERMO, M R, HOMZIAK, J and TEETER, A M (1990)
Evaluation of clamshell dredging and barge overflow, Military Ocean Terminal Sunny Point, North Carolina
Tech Rep D-90-6, US Army Engineer Waterways Experiment Station, Vicksburg, MS

PARTHENIADES, E (1965)
Erosion and deposition of cohesive soils
Journal of the Hydraulics Division, ASCE, vol 91, no HY1, pp 105–139

PEARSON, T H and ROSENBERG, R (1978)
Macrobenthic succession in relation to organic enrichment and pollution of the environment
Oceanography and Marine Biology: an Annual Review, vol 16, pp 229–311

PENNEKAMP, J G S and QUAAK, M P (1990)
Impact on the environment of turbidity caused by dredging
Terra et Aqua, no 42

PENNEKAMP, J G S, BLOKLAND, T and VERMEER, E A (1991)
Turbidity caused by dredging compared to turbidity caused by normal navigational traffic
In: *Proceedings of CEDA Dredging Days*, Amsterdam, November

PENNEKAMP J G S, and BOSLAND, P (1989)
Troebelheidsmetingen bij. Pneuma-Baggersysteem in de Berghaven te Hoek van Holland
Report J413 to Rijkswaterstaat and CSB (Bagt. 440)

PENNEKAMP J G S, EPSKAMP, R J C, ROSENBRAND W F, MULIE, A, WESSEL, G L, ARTS, T and DEIBEL, I K (1996)
Turbidity caused by dredging: viewed in perspective
Terra et Aqua, vol 64, pp 10–17

PENNEKAMP JGS, and BOSLAND P (1991)
Troebelheidsmetingen rand Milieuschijfeutter in de Berghaven te Hoek van Holland
Report J562 to Rijkswaterstaat Directie Binnwateren/RIZA

PERRY, K A (1995)
Sulphate-reducing bacteria and immobilisation of metals
Marine Georesiurces and Geotechnology, vol 13 (1–2), pp 33–39

PIANC (1996)
Handling and treatment of contaminated dredged material from ports and inland waterways. "CDM" Volume 1
Report of the Working Group 17 of the Permanent Technical Committee I Supplement to Bulletin No 89

PIANKA, E R (1970)
On r- and K-selection
American Naturalist, vol 104, pp 592–597

POINER, I R and KENNEDY, R (1984)
Complex patterns of change in the macrobenthos of a large sandbank following dredging
Marine Biology, vol 78, pp 335–352

POSFORD DUVIVIER ENVIRONMENT (1999)
Watchet Harbour Revision Order – Environmental Statement
Prepared for West Somerset District Council

POSFORD DUVIVIER ENVIRONMENT (2000)
The implications of aggregate extraction for marine SACs
In preparation for English Nature, SNH, CCW, the Environment and Heritage Service of the DOENI

REINE, K J, DICKERSON, D D and CLARKE, D G (1998)
Environmental Windows Associated with Dredging Operations. Technical Note DOER-E2
Produced by the Dredging Operations and Environmental Research Program, December, 1998

REINKING, M W (1993)
The development of special remedial dredging technique: the environmentally friendly auger dredger
In: *Proceedings of CEDA Dredging Days*, Amsterdam, November

RHOADS, D C, McCALL, P L and YINGST, J Y (1978)
Disturbance and production on the estuarine seafloor
American Scientist, vol 66, pp 577–586

ROSE, C D (1973)
Mortality of marked-sized oysters (Crassostrea virginica) in the vicinity of dredging

ROSENBERG, R (1985)
Eutrophication – the Future Marine Coastal Nuisance?
Marine Pollution Bulletin, vol 16, no 6, pp 227–231

SCHAFER, W (1972)
Ecology and Palaeoecology of Marine Environments
University of Chicago Press

SELBY, I and OOMS, K (1996)
Assessment of Offshore Sand and Gravel for Dredging
Terra et Aqua, vol 64, pp 18–28

SHAW, T J, GIESKES, J M and JAHNKE, R A (1990)
Early diagenesis in differing depositional environments: the response of transition metals in pore water
Geochimica Cosmochimica Acta, vol 58, pp 1233–1246

SHAW, T J, SHOLKOVITZ, E R and KLINKHAMMER, G (1994)
Redox dynamics in Chesapeake Bay: the effect on sediment/water uranium exchange
Geochimica Cosmochimica Acta, vol 58, pp 2985–2995

SHERK, J A (1971)
Effects of suspended and deposited sediments on estuarine organisms
Chesapeake Biological Laboratory, University of Maryland. Contribution no 443

SIMON, J L and DYER, J P III (1972)
An evaluation of siltation created by bay dredging and construction company during oyster shell dredging operations in Tampa Bay, Florida
Final Research Report, University of South Florida, Tampa

SNELGROVE, P V R and BUTMAN, C A (1994)
Animal-sediment relationships revisited: cause versus effects
Oceanography and Marine Biology: an Annual Review, vol 32, pp 111–177

SOAEFD (1996)
Monitoring and Assessment of the Marine Benthos at UK Dredged Material Disposal Sites
Scottish Fisheries Information Pamphlet No 21 1996
Prepared by the Benthos Task Team for the Marine Pollution Monitoring Management Group Coordinating Sea Disposal Monitoring for the Scottish Office Agriculture, Environment and Fisheries Department

STANDAERT, P, CLAESSENS, J, MARAIN, J and SMITS, J (1993)
The scoop dredger, a new concept for silt removal
In: *Proceedings of CEDA Dredging Days*, Amsterdam, November

SUSTAR, J F, WAKEMAN, T H and ECKER, R M (1976)
Sediment-water interaction during dredging operations
In: *Proceedings of the specialist conference on dredging and its environmental effects*, Mobile, Alabama, January

TAYLOR, J L and SALOMAN, C H (1968)
Some effects of hydraulic dredging and coastal development in Boca Ciega Bay, Florida
Fishery Bulletin, vol 67(2), pp 213–241

TEAL, C J, STEWART, D L, DEARNALEY, M P and BURT, T N (1993)
Environmental monitoring of dredging activities
In: *Proceedings of CEDA Dredging Days*, Amsterdam, November

USACE (1988a)
Sediment re-suspension by selected dredges. Environmental effects of dredging
Technical Note EEDP-09-2, USACE, WES, Vicksburg, Mississippi

USACE (1988b)
A preliminary evaluation of contaminant releases at the point of dredging. Environmental effects of dredging
Technical Note DRP-91-3, WES, Vicksburg, Mississippi

USACE (1991)
Mobile, Alabama, Field data collection project, 18 August to 2 September
Technical Report DRP-91-3. USACE, WES, Vicksburg, Mississippi

USACE (1992)
Tylers Beach, Virginia, Dredged material monitoring project, 27 September to 4 October 1992
Technical Report DRP-92-7. USACE, WES, Vicksburg, Mississippi

USAWES (1995)
The automated dredging and disposal alternatives management system (ADDAMS)
WES Environmental Laboratory Technical Note EEDP-06-12

USEPA (1976)
Impacts of construction activities in wetlands of the United States
US Environmental Protection Agency, Report EPA/600/3-76-045

USEPA (1985)
Initial mixing characteristics of municipal ocean discharges: Volume 1, Procedures and Applications
US Environmental Protection Agency, Report EPA/SW/MT-86/012a

Van DOORN, T (1988)
Dredging polluted soil with a trailing suction hopper dredger
Proceedings of CEDA Dredging Day, Environmentally acceptable methods of dredging and handling harbour and channel sediments, Hamburg, 28 September, 1988

Van MOORSEL, G W N M and WAARDENBURG, H W (1990)
Impact of gravel extraction on geomorphology and the macrobenthic community of the Klaverbank (North Sea) in 1989
Rapport Bureau Waardenburg bv, Culemborg, the Netherlands

Van RAALTE, G H and BRAY, R N (1999)
Dredging Challenged
Proceedings of CEDA Dredging Days 1999, 18–19 November 1999, Amsterdam, the Netherlands

Van RIJN, L C (1984)
Sediment Transport: Part I: bed load transport
Proceedings of the ASCE Journal of Hydraulics Division, vol 110, HY10, pp 1431–1456

Van VLIET, F J W and PENNEKAMP, JGS (1995)
Troebelheidsmetingen rond een Milieuschijfcutter op het Ketelmeer
Report J1140 to POSW

Van VLIET, F J W and BOSLAND, P (1996)
Troebelheidsmetingen rond een Cutter zuiger in de Niewe Binnehaven te Emden
Report J1035 to Combinatie Speurwerk Baggertechniek, CSB (BAGT-553)

Van VLIET, F J W and PENNEKAMP, J G S (1996)
Troebelheidsmetingen rond een Emmerbaggermolen op het Ketelmeer
Report J1286 to POSW

WALDOCK, M J, PHAIN, J E and WAITE, M E (1990)
Assessment of the environmental impacts of organotin residues from contaminated sediments
Final Report to Anglian Water Authority

WATERLOOPKUNDIG LABORATORIUM (1993)
Troebelheidsmetingen rond een hydraulische graafmachine weit een ivzierbak en een open backhoe in de circulatiekom te Wijk bij Duurstede
Report to Rijkwaterstaat and CSB (Bagt. 25)

WATERLOOPKUNDIG LABORATORIUM (1994)
Troebelheidsmetingen rond een Gemodificeerde Emmerbaggermolen in de Ze Binnehaven te Schevingen
Report JO925 to Rijkswaterstaat Directie Binnenwateren/RIZA

WEIERGANG, J (1995)
Estimation of suspended sediment concentrations based on single frequency acoustic doppler profiling
In: *Proceedings of the 14th World Dredging Congress*, Amsterdam, November

WESTERBERG, H, RONNBACK, P and FRIMANSSON, H (1996)
Effects of suspended sediments on cod egg and larvae and on the behaviour of adult herring and cod
ICES, CM 1996/E 26

WHITESIDE, P G D, OOMS, K and POSTMA, G M (1995)
Generation and decay of sediment plumes from sand dredging overflow
In: *Proceedings of 14th World Dredging Congress*, Amsterdam, November

WILSON, W B (1950)
The effects of dredging on oysters in Copano Bay, Texas
Ann. Rep. Mar. Lab. Texas Game Fish amd Oyster Comm., 1948–49, pp 1–50

WINTERWERP, J C (1999)
On the dynamics of high-concentrated mud suspensions
PhD thesis for the University of Delft

WINTERWERP, J C, UITTENBOGAARD, R E and de KOK, J M (1999)
Rapid siltation from saturated mud suspensions
Proceedings of the 5th Nearshore and Estuarine Cohesive Sediment Transport Conference, INTERCOH '98, in press

WWF (1993)
Organotin compounds
Marine Factsheet, October 1993, World Wide Fund for Nature UK

A1 Legislation and regulation

A1.1 MARINE AGGREGATE DREDGING

A1.1.1 Legislation and regulation in the UK

Licensing of marine aggregate dredging

The legislation and regulation of marine aggregate dredging in the UK is described in Section 2.1.1. The following text details additional legislation and regulation associated with marine aggregate dredging.

The Environmental Impact Assessment and Habitats (Extraction of Minerals by Marine Dredging) Regulations 2000

The Environmental Impact Assessment and Habitats (Extraction of Minerals by Marine Dredging) Regulations fulfil the Government's obligation to the EC to transpose into UK legislation the provisions of EC Directive 97/11/EC. The latter amends Directive 85/337/EEC on environmental impact assessment and EC Directive 92/43/EEC on the conservation of natural habitats and of wild fauna and flora. They will make EIA mandatory and the Secretary of State for the Environment, Transport and the Regions the competent authority in England for the statutory control of marine dredging in the public interest.

It had been initially intended that the new Regulations would apply to all of the UK and come into force on 14 March 1999. However, their implementation has been delayed by legal and constitutional difficulties associated with devolution of Scotland and Wales, some of which still remain to be resolved. Nevertheless, the Regulations are expected to be in force later in 2000.

The Town and Country Planning (Environmental Impact Assessment) (England and Wales) Regulations 1999

The Regulations introduce new statutory procedures regarding EIA, for the implementation of EC Directive 85/337/EEC, as amended by Directive 97/11/EC. The Regulations contain revised definitions of development within Schedules 1 and 2 to Regulation 2(1). Schedule 2 to Regulation 2(1) contains descriptions of extractive industries for which an EIA is required if it is likely to have significant effects on the environment by virtue of its nature, size or location, for example. Schedule 2 Table 2(c) includes *Extraction of minerals by fluvial dredging*; the Regulations' thresholds and criteria apply to all development of this kind. However, the Regulations do not include aggregate extraction by marine dredging and so are considered no further in this report.

Marine minerals guidance

The DETR is producing a new series of Marine Minerals Guidance Notes (MMGs). The first two will cover the new regulatory procedures (MMG1), and a strategy for marine aggregate dredging that seeks to provide the industry with sufficient flexibility to meet demand while minimising the effects of dredging on the environment (MMG2).

Conservation (Natural Habitats &c) Regulations 1994

These regulations transpose into UK legislation the requirements of Council Directive 92/43/EEC on the conservation of natural habitats of wild fauna and flora – commonly referred to as the Habitats Directive. Its primary objective is to promote the maintenance of biodiversity, requiring member states to maintain or restore to favourable conservation status certain rare, threatened or typical natural habitats and species. The habitats and species to be protected are listed in Annex I and II respectively to the Directive. One approach by which member states are expected to achieve the Directive's objective is through the designation and protection of a series of sites, known as Special Areas of Conservation (SACs).

Before the regulations there was no existing legislative framework for implementing the Habitats Directive in marine areas. The regulations place a duty on statutory authorities to exercise their powers in accordance with the Habitats Directive, particularly to protect European marine sites such as SACs and Special Protection Areas (SPAs).

The requirements of the Habitats Directive have been included in the Environmental Assessment and Habitats (Extraction of Minerals by Marine Dredging) Regulations 1998. It is expected that the new minerals planning guidance will include recommendations concerning the Habitats Directive and its implementation through UK regulations.

The Conservation Regulations are only relevant to aggregate dredging should a proposed or ongoing operation take place within or close to a European marine site (ie an SAC). Where this is the case, the competent authority for licensing the dredging, eg the DETR, is required to make an "appropriate assessment" of the works in terms of their potential effect on the conservation status and integrity of the European marine site. In practice, the appropriate assessment is a decision-making process, and the licence applicant should provide sufficient information to allow the competent authority to make a decision. This procedure is quite similar to the EIA process, although it is focused solely on the provisions of the Habitats Directive, ie the potential affect on the designated status of the site.

A1.1.2 Legislation and regulation overseas

United States

Offshore sand and gravel extraction is regulated in the USA by the Department of the Interior's Minerals Management Service (MMS). The responsibility of the MMS is established under Federal law, in particular the Outer Continental Shelf Lands Act and its amendments. Environmental consideration is enforced further through the National Environmental Policy Act.

Under the Marine Minerals Program, the MMS has co-operative agreements with states along the Atlantic and Gulf coasts to identify suitable deposits of sand, gravel and shell, and evaluate the feasibility of its extraction. The MMS pursues a judicious approach to managing offshore minerals extraction by ensuring that all dredging and other mining activities are environmentally sound and acceptable.

A regulatory system is in place to support all policies relating to prospecting, environmental analysis, leasing and operations. For example, regulations that govern prospecting require applicants to secure permits for any activity that could, amongst other effects, disturb aquatic life or archaeological artefacts. MMS leasing regulations specify the procedures leading up to leasing, including an option for joint Federal and state co-ordination on environmental considerations. Environmental impact statements are prepared for five-year leasing programmes. Regular inspections are undertaken to ensure compliance with lease requirements, and provide for penalties or closure of operations that violate regulations.

Europe

The following paragraphs are based on information provided in CIRIA PR68 (1999), *Marine sand and gravel in north-west Europe. A fact-finding and scoping study.*

In Denmark, the Ministry of the Environment is responsible for administering aggregate dredging. Applications for dredging permits for up to ten years are subject to a government view procedure including public and private involvement. The licence applicant has to provide sufficient information concerning the volume and quality of the mineral resource and is required to undertake an EIA.

To protect the environment and coastline, government policy in the Netherlands limits the areas from which marine aggregates can be extracted. Extraction areas have to be more than 20 m below sea level and 20 km offshore. Aggregates can only be extracted to depths of 2 m below the general seabed level, except for the approach channels to the ports of Rotterdam and Amsterdam where up to 5 m can be extracted.

In Germany, there was no official registration of aggregate dredging by local authorities until 1990. Since this date offshore aggregate extraction has been the responsibility of Das Oberbergamt. In France, the Ministry of Industry administers the application process for aggregate extraction, which involves a lengthy procedure in order to secure mining rights and permits.

A1.2 CAPITAL AND MAINTENANCE DREDGING

A1.2.1 Legislation and regulation in the UK

EIA requirements for capital and maintenance dredging

The regulations listed in Section 2.1.2 introduce new statutory procedures regarding EIAs for the implementation of EC Directive 85/337/EEC, as amended by Directive 97/11/EC. The regulations contain revised definitions of developments within Schedules 1 and 2 to Regulation 2(1). Neither Schedule makes specific reference to capital or maintenance dredging projects. However, it is possible that dredging may be subject to EIA should it fall within the project requirements of one of the following developments in Schedule 1 Parts 8(a) and 8(b):

- inland waterways and ports for inland-waterway traffic that permit the passage of vessels more than 1350 t
- trading ports, piers for loading and unloading connected to land and outside ports (excluding ferry piers) that can take vessels of more than 1350 t.

A Schedule 1 development is made subject to an EIA in accordance with the regulations since it is assumed it will significantly affect the environment because of its nature, size or location etc. Schedule 2 to Regulation 2(1) contains descriptions of infrastructure projects for which an EIA is required if it is likely to have a significant effect on the environment by virtue of factors such as its nature, size or location. Schedule 2, Table 10(g) includes *Construction of harbours and port installations including fishing harbours (unless included in Schedule 1)*, for which the regulations' thresholds and criteria apply to all development if the area of the works exceeds 1 ha.

The regulations transpose into UK legislation the requirements of the Habitats Directive (see Section A1.1.1). Like aggregate dredging, the regulations are only relevant to capital and maintenance dredging should a proposed or ongoing operation take place within or nearby a SAC or SPA.

Typically this situation arises during a dredging application under the Harbours Act or the Coast Protection Act, thereby initiating the requirement for an EIA. The competent authority for issuing statutory consent, for example the DETR, is required to make an appropriate assessment of the works in terms of the effect on the conservation status and integrity of the European marine site. In practice, the appropriate assessment is a decision-making process, and the applicant is expected to provide sufficient information to enable the competent authority to take a decision. This procedure is quite similar to the EIA process, although it focuses solely on the provisions of the Habitats Directive.

Food and Environment Protection Act 1985

The control of the deposit of dredged material at sea in the UK is administered by MAFF under Part II of the Food and Environment Protection Act (FEPA). The Act provides the framework under which MAFF controls the deposits of all wastes in the sea and provides for environmental protection consistent with national policy and international obligations arising from the London Convention 1972, Oslo Convention 1974 and the Oslo-Paris Convention 1992. While the Act gives MAFF control over the deposit of dredged material at sea, it does not give any control over dredging processes. It has been mooted that the Act should be changed to include hydrodynamic dredging, including plough dredging.

A1.2.2 Legislation and regulation overseas

Europe

All member states of the European Union are subject to the provisions made under EC directives, including EC Directive 85/337/EEC, as amended by Directive 97/11/EC. Accordingly, the need to consider the effects of sediment plumes arising from capital and maintenance dredging applies to proposed projects falling either within Annex I or to EC Directive 85/337/EEC, as amended by Directive 97/11/EC, or Annex II and is likely to have significant environmental effects.

United States

The US approach to permitting dredging and dredged material disposal is a holistic one that focuses on avoiding significant adverse effects to human health and the marine environment. The USACE and EPA are jointly responsible for permitting procedures under the Clean Water Act, Marine Protection, Research and Sanctuaries Act, Resource Conservation and Recovery Act and the Water Resources Development Act.

Hong Kong

The Environmental Impact Ordinance came into force in April 1998, superseding environmental requirements under previous legislation including the Foreshore and Seabed (Reclamation) Ordinance. It was established to avoid, minimise and control adverse environmental impacts of designated projects through the EIA process and environmental permit.

The ordinance includes a list of Schedule 2 projects that require EIA and environmental permits. The projects include reclamation works (including associated dredging) of more than 5 ha and dredging operations exceeding 50 000 m^3. Both types of project can involve capital and maintenance dredging. A permit for smaller dredging operations is required where a project is less than 500 m from environmentally sensitive site (eg a marine park or reserve, bathing beach, fish culture zone) or is less than 100 m from a seawater intake point. The distances contained within the project descriptions suggest that Hong Kong's legislation recognises that dredging-induced sediment plumes can affect the environment.

To date, the Environmental Protection Department has issued a permit for dredging about 3 550 000 m^3 of seabed for relocating mooring buoys and two permits for dredging projects (130 000–150 000 m^3) within fish culture zones. All three permits include detailed monitoring and auditing conditions on dredging operations to prevent adverse environmental effects associated with the introduction of sediment into the water column.

A2 Sediment plume stakeholders

A2.1 THE DREDGING INDUSTRY

A2.1.1 Dredging companies and contractors

It is important to note that many dredging companies and contractors operate dredging plant able to carry out both aggregate dredging and capital and maintenance dredging. For example, large trailing suction hopper dredgers are used for winning aggregates for land-based projects as well as for coastal reclamation projects (ie capital works).

In the UK, the extraction of marine aggregates tends to be undertaken by UK companies that solely supply the land-based construction industry, whereas multinational dredging contractors tend to dominate capital and maintenance dredging for ports and other maritime projects. The situation is different in Europe, where aggregate dredging and capital and maintenance dredging are not separate industries.

As a result, the UK aggregate dredging industry tends to comprise private British companies rather than multinational dredging contractors based in the UK, whereas in Europe multinational dredging contractors undertake all dredging projects.

A2.1.2 Dredging associations

British Marine Aggregate Producers' Association

BMAPA was established to increase awareness of the marine aggregate dredging industry and to raise the profile and promote the interests of the aggregate dredging industry. It also acts as a focal point for liaison with other coastal and marine user groups. BMAPA is affiliated with the Quarry Products Association (QPA).

Federation of Dredging Contractors

The membership of this trade association for major dredging contractors is open to contractors with an operating office in the UK and the capability to work internationally. The Federation represents members' interests on matters of common concern.

International Association of Dredging Contractors

The IADC, based in the Netherlands, is a worldwide umbrella organisation for more than 120 private dredging companies. The IADC seeks to promote the private dredging industry and to establish fair and open market conditions for its members. The IADC promotes dredging to its members and the public through the development of standard contracts and conditions, conference organisation, publications and sponsoring research.

Central Dredging Association

CEDA provides a forum for individuals and organisations involved in dredging, particularly those in Europe, Africa and the Middle East. CEDA is part of the World Organisation of Dredging Associations (WODA), which also comprises the Western Dredging Association (WEDA) covering North, Central and South America, and the

Eastern Dredging Association (EADA) covering the Indian subcontinent, Asia and Australasia.

CEDA is based in the Netherlands and has British, Belgian and Dutch national sections. Like the other associations, it aims to promote education about dredging, disseminate quality information on dredging, develop guidelines and standards relating to good practice, initiate and support research, and be proactive in relevant policy-making. CEDA seeks to accomplish its aims through providing a multidisciplinary forum, organising conferences and seminars, publishing reports and other material, arranging lectures and site visits, and acting as official observers at international conventions.

Vereniging van Waterbouwers in Bagger-, Kust- en Oeverwerken

The VBKO is the Dutch association of contractors that undertake dredging, shore and bank protection works, representing more than 300 companies. Its aim is to promote its members' interests where affected by politics, economics, legislation, social affairs and technical affairs, and provides an outlet for education and business promotion through Dutch and international contacts. Recently, Dutch dredging companies and the Rijkswaterstaat, through the VBKO, have commissioned UK investigations into turbidity assessment software.

Permanent International Association of Navigational Congresses

PIANC, based in Belgium, represents organisations involved in the safe and efficient operation of all types of commercial and recreational vessels. Its mission covers the promotion, management and sustainable development of navigational inland, coastal and ocean waterways, including ports and harbours, logistics, infrastructure and the coastal zone.

A2.2 REGULATORS

A2.2.1 Government offices

Aggregate dredging licensing authorities

The new regulations described in Section 2.1.1 define as regulators the Secretary of State for the Environment, Transport and the Regions, Scottish ministers, the National Assembly for Wales and the DOENI. They have responsibility for the statutory control of marine aggregate dredging within the waters of, respectively, England, Scotland, Wales and Northern Ireland.

Capital and maintenance dredging licensing authorities

Capital dredging is subject to a statutory order or consent under Harbours Act 1964 or the Coast Protection Act 1949, depending on the circumstances of a specific dredging project. In the future, it is likely that the Secretary of State for the Environment, Transport and the Regions, Scottish Ministers, the National Assembly for Wales and the DOENI will be responsible for regulating such matters.

A2.2.2 Ministry of Agriculture, Fisheries and Food

Aggregate dredging

MAFF is responsible for advising the DETR on issues such as marine biology and fisheries. For these purposes, MAFF takes advice from CEFAS and the SFI and consults fishing organisations.

CEFAS is an executive agency of MAFF and was formerly the Directorate of Fisheries Research. CEFAS advises MAFF on the environmental effects of aggregate dredging and on the disposal of dredged material at sea and marine construction. The SFI advises MAFF about issues regarding commercial fisheries interests. It liaises with the various fishing organisations including sea fisheries committees and fishermen's organisations.

Capital and maintenance dredging

Although the Food and Environment Protection Act 1985 gives MAFF control over the deposit of dredged material at sea, it does not give any control over dredging processes. The Act might be changed to include hydrodynamic dredging, including plough dredging, but this change is likely to focus on the sedimentation (ie disposal) effects of hydrodynamic dredging, rather than the effects of the plume in the water column.

A2.2.3 Conservation agencies

The government's advisers on nature conservation issues are English Nature, the Countryside Council for Wales (CCW), Scottish Natural Heritage (SNH) and the Department of the Environment for Northern Ireland (DOENI). These agencies are responsible for advising the government offices about the potential consequences of various proposals, including dredging projects, where statutory requirements might be compromised. In the case of dredging, one of the conservation agencies' main concerns is the protection of marine biodiversity, as required under the EC Habitats Directive (implemented in the UK under the Conservation (Natural Habitats etc) Regulations 1994). In particular, the conservation agencies advise the government offices on the potential effects of aggregate dredging on marine SACs and the effects of capital and maintenance dredging on SACs and SPAs.

A2.3 OTHER INTERESTED PARTIES

Port and harbour authorities

Port and harbour authorities are the statutory undertakers responsible for providing safe navigation into the areas under their control. This provision often requires these authorities to carry out maintenance dredging. Their powers are confirmed under Acts of Parliament or harbour empowerment/revision orders granted under the Harbours Act 1964. Port and harbour authorities include privately owned port companies and terminal operators, local authorities and trust ports.

Fishery interests

In addition to the SFI within MAFF, fisheries interests are represented by regional sea fisheries committees (eg Southern Sea Fisheries Committee), the Environment Agency, shellfishery organisations (eg Shellfish Association of Great Britain) and local fishermen's associations (eg Teign Musselmen's Society). The Environment Agency is responsible for looking after the statutory interests associated with the migration of freshwater and estuarine fish and shellfish species including migratory fish (eg salmon and trout).

Crown Estate

The Crown Estate owns around 55 per cent of the UK's foreshore and the seabed out to the 12-mile (19 km) territorial limit. It also owns the rights to explore and exploit the natural resources of the UK's continental shelf, including marine aggregates, but excluding oil, gas and coal. Until recently, the Crown Estate issued prospecting and extraction licences following an informal government view procedure. The new procedures described in Section A1.1.1 has replaced this practice.

Environmental specialists and consultants

Specialists and consultants of various types carry out research and investigations into the plumes arising from dredging, including private consultancies, public agencies, government-sponsored research establishments and universities. In particular, these organisations are concerned with predicting the mechanisms for the generation and transport of sediment plumes and the environmental effects associated with them. Most of these organisations provide commercial services to the dredging industry and/or regulators and other interested parties.

A2.4 KNOWLEDGE TRANSFER

A2.4.1 Knowledge transfer in the UK

Dredging companies and contractors

There tends to be little direct knowledge transfer between dredging companies since this might affect commercial competitiveness. However, several dredging industry associations serve as focal points for information gathering and knowledge transfer.

Dredging associations

The dredging industry informs its members and the public about dredging through international organisations such as the IADC, CEDA, PIANC and the FDC. They carry out their activities independently or in co-operation with other dredging and maritime organisations. For example, the IADC adopts a number of approaches to disseminating information, including the publication of books and magazines (eg *Terra et aqua*), attendance at and organisation of conferences and the sponsorship of dredging seminars. It is these types of activities that bring together experts in the subject of dredging and present focal points for the transfer of knowledge on issues concerning dredging, including issues associated with sediment plumes. The dredging associations also co-operate on various projects, for example, CEDA and the IADC are jointly issuing a series of booklets collectively entitled *Environmental aspects of dredging*.

The dredging industry is aware that the public and many potentially important decision-makers, are unaware of the nature of dredging operations. In response, a group of dredging organisations, comprising the IADC, IAPH, PIANC and WODA, has recently published *Dredging: the facts*. The purpose of this document is noted in an accompanying leaflet, which states: "Proposed dredging or disposal projects occasionally run into problems, and objections may be raised. Experience has demonstrated that many such objections to a proposal to dredge or dispose of dredged materials are, in fact, based on an incomplete understanding of the proposed operation and its impacts. People (decision-makers and the public alike) who have not previously have had any reason to be aware of, or become involved in, dredging projects may not understand what a dredger is or how it operates, or even why dredging is necessary."

Regulators

In the UK, the regulators do not have a formal programme for the release of detailed information and research, including information on sediment plumes arising from dredging. For example, CEFAS makes available information on an *ad hoc* basis to enquirers and reports on its research through international fora, including the ICES sand and gravel working group.

Conservation agencies

The importance of knowledge transfer is acknowledged by conservation agencies under the UK Marine SACs Project. The project identified seven categories of human activity – one of which was aggregate extraction – that have the potential to affect marine areas of European nature conservation value. The project recognised that knowledge on the activities was extensive but widely dispersed, introducing a requirement for collation.

Box A2.1   *The UK Marine SACs Project*

> The UK Marine SACs Project involves a four-year partnership (1996–2001) between English Nature, SNH, the CCW, the Environment and Heritage Service of the DOENI, the JNCC and the SAMS. The project's overall objective is to facilitate establishment of management schemes for 12 candidate marine SACs. One of the main components of this project is to assess the interactions that can take place between human activities and the habitat and species interests at these sites.
>
> The preface to the outputs from the initiative summarises the reasons why knowledge transfer is important. "The reports are aimed primarily at staff from the relevant authorities who jointly have the responsibility for assessing activities on marine SACs and developing the appropriate set of management measures to ensure the marine features remain in favourable condition. But industry, user and interest groups have an important role too in advising relevant authorities in these decisions and so the reports will provide a valuable resource for them. Finally, as the knowledge and learning from this Project will be applicable to other areas beyond the UK, so these reports have a role in guiding practitioners elsewhere in Europe."

A2.4.2 Knowledge transfer overseas

Europe

The Marine Sand and Gravel Information Service (MAGIS) is a relatively new approach to providing specialist information for use by anyone involved or interested in the European marine aggregate industry. Set up through the EU's INFO 2000 programme, which provides funding for data support to all business sectors, MAGIS involves dredging contractors, government licensing agencies and dredger manufacturers.

Government agencies from the UK, the Netherlands, Belgium and Denmark participate in MAGIS and it is intended that their involvement should improve harmonisation of aggregate extraction licensing and data management. In addition, associations including the IADC, CEDA and the VBKO represent the dredging industry. Although current

MAGIS literature and website do not cover sediment plumes within their itineraries, data and information is updated bi-weekly on line and in a newsletter. In the future, MAGIS offers much potential for disseminating information about sediment plumes to stakeholders within the aggregate dredging industry throughout Europe.

International dredging associations, such as the European-based IADC and CEDA, transfer knowledge about dredging through the organisation of conferences and seminars. For example, CEDA annually holds its Dredging Days to coincide with the Europort Exhibition in the Netherlands. In addition, these associations promote the dredging industry and provide technical information about dredging projects and new research through the publication of journals such as *Terra et aqua* and guidance documents including the joint IADC/CEDA series of seven publications entitled *The environmental aspects of dredging*.

Perhaps the Internet is the most efficient method of sharing data, including information about sediment plumes. Access is restricted only by the availability of the computing hardware and software necessary to get on line via a telephone link to an Internet server provider. A major advantage of the Internet is its freedom from corporate regulation, such that most information can be downloaded or printed off the Internet free of cost.

United States

The United States extensively uses the Internet to disseminate government information, an approach accelerated by statutory duties under the Freedom of Information Act. In terms of dredging and the sediment plumes associated with dredging, two US government agencies in particular provide vast amounts of publicly accessible information through the Internet:

- the Engineer Research and Development Centre of the US Army Corps of Engineers (USACE)
- Minerals Management Service of the US Department of the Interior.

A3   Field measurements of losses from dredging operations

This appendix summarises published field data. It is not intended to be a complete resource of measurements of sediment losses arising from plumes, which is beyond the scope of this book. However, it is intended to provide a representative sample of the available measurements and to describe key measurements in detail where appropriate. It has not always been possible to detail the conditions under which measurements were made as such information was often not recorded by the authors of the original source.

Caution

The following sections summarise published field measurements of plume concentrations arising from different dredging techniques. As stated at the start of Section 3.3.1 although the current state of knowledge is effectively based on field measurement data the general applicability of the reported data is limited and care should be taken in using it to draw conclusions about the effects of different types of dredging activity.

A3.1   TRAILING SUCTION HOPPER DREDGING MEASUREMENTS

Capital and maintenance dredging

Blokland (1986) and Blokland *et al* (1986) describe measurements taken around maintenance trailing suction hopper operations in silty muds. These measurements are summarised in Table A3.1 together with measurements from maintenance dredging in Grays Harbour, Washington. The measurements indicate that the losses (S-factors) associated with a number trailing suction hopper maintenance dredging are of the order of 1–13 kg/m^3 and increase with the amount of overflow. These losses compare will the range of values given in Table 3.2 in the main text (Kirby and Land, 1991).

Observations at Grays Harbour showed that in the case of dredging with no overflow, turbidity increases were limited to the proximity of the bed. The presence of overflow produced a plume behind the dredger that was seen to settle rapidly to the bed. These observations indicate the importance of the dynamic plume phase in the behaviour of overflow released from the dredging vessel.

The large variation between apparently identical dredging operations at Buitenhaven, Delftzijl, shows that current magnitudes, dredging operating speed and spatial variations in the type of material being dredged can together greatly affect the extent of losses of material from trailing suction dredging operations.

Table A3.1 *Measurements of turbidity around a trailing suction hopper dredger*

CAUTION SHOULD BE TAKEN IN USING THESE DATA (see Section 3.3.1)

Location	3e Petroleumhaven, Rotterdam		Buitenhaven, Delfzijl		Gray's Harbour, Washington	
Date	29/05/85	29/05/85	15/04/86	16/04/86	N/A	N/A
Soil composition	58 % <16 μm 5 % >63 μm	58 % <16 μm 5 % >63 μm	74 % <16 μm 10 % >63 μm	74 % <16 μm 10 % >63 μm	N/A	N/A
In situ density (kg/m^3)	1170 top 0.6 m 1300 below	1170 top 0.25 m 1300 below	1100 top 0.3 m 1300 next 1.5 m	1100 top 0.3 m 1300 next 1.5 m	N/A	N/A
Dredging technique	no overflow	overflow	little overflow	little overflow	with overflow	without overflow
Trail length	200 m	200 m	100 m	100 m	N/A	N/A
Hopper volume	6100 m^3	6100 m^3	803 m^3	803 m^3	N/A	N/A
Production (m^3/h)	5400	4125	1750	1750	N/A	N/A
Current velocity	none	none	none	0.2 m/s	N/A	N/A
Water depth	13 m	13 m	9 m	9 m	N/A	N/A
Background concentration	40 ppm	75 ppm	60 ppm	70 ppm	28–60 ppm	12–54 ppm
Turbidity increase	150 ppm	400 ppm	10 ppm	20 ppm	surf: 100 ppm bed: 700 ppm	surf: negligible bed: 40–50 ppm
Duration of turbidity increase	1 hr	1hr	0.5 hr	1 hr	N/A	N/A
Re-suspension of sediment	4 kg/m^3	13 kg/m^3	1 kg/m^3	5 kg/m^3	N/A	N/A

Table A3.2 shows measured concentrations along the centrelines of the paths of trailing suction dredgers dredging with overflow at Mare Island Strait, Richmond Harbor and Alameda Naval Air Station in the USA (Sustar *et al*, 1976). Higher levels (more than 5000 ppm) were observed adjacent to the dredger. However, such concentrations reduce rapidly to give the values shown in Table A3.2 (ie peak plume centreline values of 190–1100 mg/l and average plume centreline values of 40–700 mg/l). Details of the conditions under which measurements were made are unavailable.

Table A3.2 *Typical sediment disturbances around a trailing hopper dredge*

CAUTION SHOULD BE TAKEN IN USING THESE DATA (see Section 3.3.1)

Project	Depth (m)	Background (ppm)	Centreline		50 m off centreline		100 m off centreline	
			max (ppm)	avg (ppm)	max (ppm)	avg (ppm)	max (ppm)	avg (ppm)
Mare Island Strait	1	33	210	210	60	43	12	12
	5	83	110	64	46	46	49	49
	10	123	1110	743	2600	337	260	233
Richmond Harbor	1	31	82	65	51	45	23	23
	5	33	39	33	55	55	32	32
	10	39	200	145	–	–	–	–
Alameda Naval Air Station	1	35	188	131	–	–	–	–
	5	28	47	42	–	–	–	–
	10	38	58	58	–	–	–	–

Specific on-board field measurements have been carried out to examine the re-suspension of fine sediment during aggregate dredging in Hong Kong, and these have been reported by Whiteside *et al* (1995). The paper considers the loading of the 8225 m^3 trailing dredger HAM 310 over a period of 90 minutes. It was shown that for a total measured load of 11 500 t the total overflow loss was 3000 t. It was estimated that approximately 50 per cent of the overflow was of fine particles (<0.063 mm) and the average rate of overflow of fine material over the 90 minute period was 280 kg/s.

Acoustic measurements and water sampling were undertaken during dredging of silt and clay by two different dredgers at Lough Foyle in February 1993. The results are summarised in Table A3.3 below.

Table A3.3 *Trailing suction hopper measurements at Lough Foyle*

CAUTION SHOULD BE TAKEN IN USING THESE DATA (see Section 3.3.1)

	Trial 1	Trial 2	Trial 3	Trial 4
Dredger name	Cornelia	Cornelia	WD Medway II	WD Medway II
Site name	Greenbank Light area	Kilderry Light area	Longfield Light area	Lisahally turning circle
Water depth	7 m	9.5 m	10 m	5–6 m
Max conc increase	90 mg/l close to bed	2500 mg/l close to bed	410 mg/l close to bed	585 mg/l in plume
Background conc	190 mg/l	540 mg/l	130 mg/l	30 mg/l

Aggregate dredging

In 1995 HR Wallingford, Coastline Surveys and the main UK Aggregate Dredging Contractors carried out a dedicated field survey in the English Channel using acoustic profilers and water sampling techniques (HR Wallingford, 1996b). The plumes resulting from aggregate dredging operations on the Owers Bank for three vessels: of 1300 m^3 hopper size, overflowing via spillways and dredging at anchor; of 1300 m^3, overflowing via spillways and dredging while under way; and of 7500 m^3, overflowing via a central spillway discharging through the hull. Measured concentrations of muddy material (< 0.063 mm in diameter) varied up to 29 mg/l, 25 mg/l and 209 mg/l respectively, with concentrations reducing to background levels (of the order of 8 mg/l) 100 m to 500 m from the point of dredging. Concentration increases caused by sandy material (>0.063 mm in diameter) varied up to 611 mg, 1200 mg/l and 2117 mg/l, respectively, near the bed. The material dredged was gravelly sand with a small amount of mud and water depths varied between 18.5–21 m. Overflow concentrations and particle size distribution were measured during dredging of the largest vessel. On average the totals solids concentration of the overflow was 5500 mg/l, of which 65 per cent was silt (diameter < 0.063 mm) and 10 per cent fine sand (0.063 mm < diameter < 0.125 mm). Currents varied over the range 0.2–0.7 m/s.

Samples were taken to measure the quantity of solids being lost from the overflow spillways and reject chute during aggregate dredging 20 km off the east coast of the UK between Great Yarmouth and Lowestoft (HR Wallingford, 1994b). The bed material at the trial location comprised sandy gravel on or just below the seabed, with up to 5 per cent silt and 10 per cent fine sand, respectively. Screens were used to divert water plus a proportion of material less than 5 mm in diameter overboard; the resulting load was 60 per cent stone and 40 per cent sand. The study found that the release of fine material via the screening chute was more than ten times that released by the spillways and the combined loss rate of fines to the water column was about 20 kg/s over a five-hour period. Current speeds and the extent of wave action during the field exercise were not recorded.

Losses from a trailing suction hopper dredger were measured during aggregate dredging of an all-in load at Shingle Bank, Hastings (HR Wallingford, 1993b). This study concluded that 11 kg total dry solids per m^3 of overspill passed over the spillways during a 2.5-hour dredging period when a total mass of 4400 t of dredged material was recovered. Of this total, 4150 t were retained and 250 t were washed out during dredging (130 t of silt/clay and 120 t of sand). The average rate of input of fines to the water column in this case was about 14 kg/s. The bed material at Shingle Bank Hastings is up to 4 per cent silt. Current speeds and the extent of wave action during the field exercise were not recorded.

The following table summarises measurements of loss rates made on behalf of a consortium of aggregate dredging companies in the UK (MMS, 1998). The values relate to measurements on board the dredger rather than in-situ in the surrounding waters.

Table A3.4 *Summary of UK spillway measurement losses*

CAUTION SHOULD BE TAKEN IN USING THESE DATA (see Section 3.3.1)

Material size (mm)	Rate* of input to water column (kg/s)
<0.063	14–20
0.063–0.125	2–10
0.125–0.250	5–40
0.250–0.500	5–136
0.500–1.000	1–118
>1.000	0–135

*   The lower of the two quoted values applies to the case of loading all-in, the higher value corresponds to the case of screening.

A3.2 CUTTER SUCTION MEASUREMENTS

Measurements undertaken during cutter suction operations are described below. There is some evidence to suggest that concentration increases associated with cutter suction operations are sensitive to distance from the cutter head and such data has not always been recorded by field scientists.

Collins (1995) analysed USACE data from three sites to derive loss rates. The site records and the soils data were very poor, making the loss estimates of doubtful usefulness. The loss rates varied from 0.013 kg/m^3 to 4.4 kg/m^3 for cutterhead diameters of 0.9–1.8 m and sediments varying from very soft clay to silty loam. One of these sites, Calumet Harbor, recorded increases of turbidity of only 2–5 mg/l, while another, James River, recorded increases of 150–350 mg/l. The natures of the sediment and current conditions are shown in Table A3.5. Comparison of the cutter suction dredging measurements with the grab dredging measurements at Calumet Harbor presented in Table A3.7 suggests that the losses associated with cutter section dredging are much lower than with grab dredging.

Table A3.5 *Measurements of turbidity around a cutter suction dredger*

CAUTION SHOULD BE TAKEN IN USING THESE DATA (see Section 3.3.1)

Location	Calumet Harbor	James River
Soil composition	Soft organic clay/silt 80 per cent fines	Silty clay
In situ density	Not recorded	Not recorded
Trail length	Not recorded	Not recorded
Hopper volume	Not recorded	Not recorded
Production rate	Not recorded	Not recorded
Current velocity	0.0–0.05 m/s	0.15–0.7 m/s
Water depth	Not recorded	Not recorded
Background concentration	Surface 2 mg/l Bottom 5 mg/l	Surface 42 mg/l Bottom 86 mg/l
Turbidity increase	× 2.0	× 3.8
Duration of turbidity increase	Not recorded	Not recorded

Measurements of suspended sediment concentrations were undertaken during dredging of silt and clay by the SEINE at Nieuwe Binnenhaven, Emden, in October 1995 (Van Vliet and Bosland, 1996). The dredging was carried out in 9–11 m depth of water with a six-bladed cutterhead of 1.4 m radius. The material dredged was 90 per cent silt and had an *in situ* density of 1150 kg/m^3. The results are summarised in Table A3.6 below.

Table A3.6 *Cutter suction dredging measurements at Nieuwe Binnenhaven*

CAUTION SHOULD BE TAKEN IN USING THESE DATA (see Section 3.3.1)

	Trial 1	Trial 2	Trial 3
Trial description	Cutter submerged	Cutter ½-submerged	Cutter ½-submerged
Swing speed	15 m/min	15 m/min	15 m/min
Cutter speed	20 rpm	20 rpm	20 rpm
Pipe discharge	1935 m^3/h	2355 m^3/h	2260 m^3/h
Silt screen	No	No	No
Gross production rate	710 m^3/h	470 m^3/h	600 m^3/h
Mass re-suspended	0 kg/m^3	0 kg/m^3	Not measured
Conc increase	0 mg/l	0 mg/l	Not measured
Background conc	110 mg/l	115 mg/l	145 mg/l

Crude estimates of loss rates were made based on data from the Hong Kong Contaminated Spoil Management study (Mott McDonald, 1991). The estimated S-parameters varied from 1.8 kg/m^3 to 5.8 kg/m^3 and the loss rates from 0.9 kg/s to 1.6 kg/s for silty clay and clay sediments.

The observations of re-suspension from cutter suction dredging have shown S-factors to vary over the range 0–8 kg/m^3 with corresponding loss fates of 0–1.3 kg/s. These results are of the same order as those of Table 3.2 but vary more widely, indicating the practical problems of undertaking sensitive measurements of suspended sediment concentrations close to the point of dredging. Table 3.2 indicates that losses into the water column cab be reduced by limiting the swing and rotation speed of the cutter, which has been established by laboratory research (eg Brahme and Herbich, 1986, Andrassay and Herbich, 1988).

A3.3 GRAB DREDGING MEASUREMENTS

Measurements were undertaken during the following operations, briefly described here and summarised in Tables A3.7 and A3.8.

At Nieuwekerk aan de IJssel, the Netherlands, Pennekamp and Quaak (1990) recorded three sets of measurements around grab dredging operations in 5 m water depth using an open 2.5 m^3 grab and a closed 3 m^3 grab. Two tests were carried out within an enclosed silt screen.

Open grab resulted in losses three times greater than the closed grab and losses were (probably) reduced by a factor of 6 by the silt screen. S-parameters ranged from 5 kg/m^3 (closed grab with silt screen) to 20 kg/m^3 (open grab, no silt screen). These measurements illustrate the effectiveness of watertight grabs and silt screens in reducing the loss of fine sediment into the water column.

At Merwedehaven, Rotterdam, Pennekamp and Quaak (1990) recorded measurements around a 1.1 m³ working grab working in 11 m of water. No silt screen was used. The S-parameter was 3 kg/m³.

At Oude Haven, 't Sas at Zierikzee, the Netherlands, Pennekamp *et al* (1991) recorded two sets of measurements both affected by nearby shipping movements. The first measurement was of a 2.5 m³ watertight grab and the second of a 1.3 m³ watertight grab. In both cases debris prevented proper grab closure, and S-parameters were 11 kg/m³.

Measurements were made by USACE at three grab-dredging sites: Black Rock Harbor and Duwamish Waterway (USACE, 1988a) and Calumet River (USACE, 1988b). In all cases very poor records of dredging process and soil type were obtained and results must be treated with caution. The grabs were open and varied between 7.6 m³ and 9.2 m³. Silt curtains were not used. Collins (1995) made a later attempt to estimate the release rates corresponding to these operations and sediment release rates of 0.24–1.68 kg/s were estimated. The turbidity increases recorded suggest a relatively high release of sediment.

More extensive measurements were made during grab dredging operations at the Almeda Naval Air Base Station (Sustar *et al*, 1976). These found concentrations in the centre of the resulting plume to vary up to 170 mg/l with average values of 30–90 mg/l, reducing to 30 mg/l or less at a distance of 100 m or less. Background concentrations were measured at 24–37 mg/l.

Tables A3.7 and A3.8 indicate that measured concentration increases due to grab dredging operations vary over the range 20–140 ppm with corresponding S-factors of 2–30 kg/m³. These compare well with the S-factors presented by Kirby and Land (1991) (Table 3.2). The results of field measurements also indicate the effectiveness of silt screening and watertight grabs in reducing the loss of sediment into surrounding waters.

Table A3.7 *Dutch measurements of turbidity around a grab dredger*

CAUTION SHOULD BE TAKEN IN USING THESE DATA (see Section 3.3.1)

Location	Merwedehaven Rotterdam	Nieuwekerk aan de IJssel			't Sas at Zierikzee, Eastern Scheldt	
Date	12/08/86	19/05/88			1990	
Soil composition	54 % <16 µm 8 % >63 µm	40 % <16 µm 23 % >63 µm			Silt (75 %) /clay (48 %)	
In situ density (kg/m³)	1145 top 0.6 m 1300 below	1360 top 0.25 m 1560 next m			1330 kg/m³	
Dredging technique	open 1.1 m³ grab, no silt screen	watertight 3 m³ grab with silt screen	open 2.5 m³ grab with silt screen	watertight 3 m³ grab, no silt screen	self-propelled hopper grab, watertight grab, silt screen	self-propelled hopper grab, watertight grab
Production (m³/h)	90	102	84	166	102 m³/h	88 m³/h
Current velocity	0.05 m/s	0.1 m/s	0.2 m/s	0.04 m/s	negligible	negligible
Water depth	11 m	5 m	5 m	5 m	3–4m	3–4 m
Background concentn.	20 mg/l	35 mg/l	35 mg/l	35 mg/l	50 mg/l	50 mg/l
Turbidity increase	35 mg/l	20 mg/l	35 mg/l	100 mg/l	10 5mg/l	90 mg/l
Duration of turbidity increase	1 hr	1 hr	1 hr	1 hr	NOT RECORDED	NOT RECORDED
Re-suspension of sediment	3 kg/m³	5 kg/m³	10 kg/m³	20 kg/m³	11 kg/m³	11 kg/m³

Acoustic profiler transects and water sampling were undertaken during grab dredging by the WD Tyne dredger in Lough Foyle in February 1993. Water depths were of the order of 5–10 m and the material dredged was silt and clay. Two trails were carried out for background concentrations of 45 mg/l. The first recorded a concentration increase of 20 mg/l and the second recorded an increase of 1080 mg/l close to the bed.

Table A3.8 US measurements of turbidity around a grab dredger

CAUTION SHOULD BE TAKEN IN USING THESE DATA (see Section 3.3.1)

Location	Black Rock Harbor	Duwamish Waterway	Calumet River
Date	NOT RECORDED	NOT RECORDED	NOT RECORDED
Soil composition	Sandy organic clay, 90 % fines	Sandy clayey silt	NOT RECORDED
In situ density (kg/m^3)	NOT RECORDED	NOT RECORDED	NOT RECORDED
Dredging technique	open clamshell (10 yd^3)	open clamshell	open clamshell
Production (m^3/h)	NOT RECORDED	NOT RECORDED	NOT RECORDED
Current velocity	0.05–0.25 m/s	0.1–0.3 m/s	NOT RECORDED
Water depth	NOT RECORDED	NOT RECORDED	NOT RECORDED
Background concentration	surface: 45 ppm bottom: 69 ppm	surf: 11 ppm bed: 26 ppm	10–12 ppm
Turbidity increase	× 15.9	× 6.1	surf: 40 ppm bed: 140 ppm
Duration of turbidity increase	NOT RECORDED	NOT RECORDED	NOT RECORDED
Re-suspension of sediment	NOT RECORDED	NOT RECORDED	NOT RECORDED

A3.4 BACKHOE DREDGING MEASUREMENTS

Backhoe dredging was carried out on the Amsterdam-Rhine Canal by Wijk bij Duurstede (Waterloopkundig Laboratium WL, 1993). Measurements of a backhoe with an open bucket gave an S-parameter of 54 kg/m^3. Use of a closed visor backhoe gave an S-parameter of 21 kg/m^3. These losses are comparatively high compared with other dredging methods but still indicate the effectiveness of existing techniques for reducing losses into the water column.

A3.5 BUCKET LADDER DREDGING MEASUREMENTS

There are few published measurements undertaken during bucket ladder operations but those available are described below.

Barnard (1978) described a typical bucket ladder dredge operation as producing a plume some 300 m (at the surface) to 500 m (at the bed) long with average surface concentrations of 100 mg/l or less, and maximum surface concentrations of 500 mg/l. The near-bottom concentrations were not recorded but are expected to have been higher.

Measurements reported by Gemeentewerken Rotterdam (1988) at the Noordzeekannal for silty sand showed a turbidity increase of 110 mg/l and a corresponding S-factor of 20 kg/m^3. The details of the conditions in which the measurments were made are shown in Table A3.9.

Table A3.9 *Measurements of turbidity around a bucket ladder dredger*

CAUTION SHOULD BE TAKEN IN USING THESE DATA (see Section 3.3.1)

Location	Noordzeekanal, Amsterdam
Date	25/1//87
Soil composition	65 % < 16 µm, 16 % > 63 µm
In situ density (kg/m^3)	1100 top 0.25 m, 1300 next 1 m
Dredging technique	0.7 m^3 buckets
Production (m^3/hr)	714
Current velocity	0.06 m/s
Water depth	14 m
Background concentration	15 mg/l
Turbidity increase	110 mg/l
Duration of turbidity increase	1 hour
Re-suspension of sediment	20 kg/m^3

S-parameter measurements around the SATURN (Pennekamp *et al*, 1996) gave a value of 18–21 kg/m^3 in silt. This corresponds to a loss rate of 4 kg/s.

Pennekamp *et al* (Waterloopkundig Laboratorium WL, 1994) give S-parameter measurements around the Aalscholver of 3–5 kg/m^3, which corresponds to a loss rate of the order of 0.33 kg/s. The dredger was adapted for "environmental efficiency" with an enclosed ladder and air valves fitted to the buckets.

The field measurements presented here suggest typical loss rates for bucket ladder dredging corresponding to S-factor of 20 kg/m^3. However, by taking steps to reduce losses such as enclosed ladders and air valves the data suggests losses can be reduced greatly to values of the order of 5 kg/m^3.

Measurements were taken during dredging in the Ketelmeer Lake in the Netherlands (Van Vliet and Pennekamp, 1996). The measurements, which were taken at a distance of less than 50 m from the dredger, showed depth-averaged increases in suspended sediment concentration of up to 35 mg/l with associated S-factor losses of 9 kg/m^3 and 12 kg/m^3.

A3.6 MEASUREMENTS FROM OTHER DREDGERS

Table A3.10, A3.11 and A3.12 summarise measurements taken from dredgers other than those considered above – disc cutter, pneuma, and wormwheel suction dredgers. The disc cutter and pneuma measurements were undertaken in 1990 at Berghaven, Holland, while the wormwheel suction measurements were made in Geulhaven, Holland in 1991.

Table A3.10 *Measurements of turbidity around a disc cutter dredger (Pennekamp and Bosland, 1991)*

CAUTION SHOULD BE TAKEN IN USING THIS DATA (see Section 3.3.1)

Environmental disc cutter dredger (2 m diameter)

Location	Berghaven, Hook of Holland	
Date	November 1990	
Barge/hopper	Rosalian	
Material type	Silt (17 %)/clay (80 %)	
In-situ density	1.25 kg/m^3	
Dredged density	1.185 kg/m^3	
Number of trials	2	
Measurements	Concentration profiles around operation	
Water depth	5 m	
	Trial 1	Trial 2
Silt screen	No	No
Production rate	108 m^3/h	126 m^3/h
Mass re-suspended	Not detectable	Not detectable
Conc increase	Not detectable	Not detectable
Background conc	25 mg/l	25 mg/l

Measurements were taken during dredging by an environmental disc cutter in the Ketelmeer lake in the Netherlands (Van Vliet and Pennekamp, 1995). Measurements were obtained within a distance of 50 m from the dredger and showed depth-averaged increases in suspended sediment concentration of up to 7 mg/l with associated S-factor losses of up to 1 kg/m^3.

Table A3.11 *Measurements of turbidity around a pneuma pump dredger (Pennekamp and Bosland, 1989)*

CAUTION SHOULD BE TAKEN IN USING THIS DATA (see Section 3.3.1)

Pneuma pump				
Location	Berghaven, Hook of Holland			
Date	November 1988			
Dredger type	Pneuma Dredge Pump			
Barge/hopper	de Reiger			
Material type	Silt (30 %)/clay (45 %)/sand (20 %)			
In-situ density	1.31 kg/m^3			
Dredged density	1.14 kg/m^3			
Number of trials	4			
Measurements	Concentration profiles around operation			
Water depth	5 m			
	Trial 1	Trial 2	Trial 3	Trial 4
Trial description	Fixed speed 1.5 m/min	Variable 4.75–6.5 m/min	Variable 0.5–2.0 m/min	Variable speed and depth
Silt screen	No	No	No	No
Gross production rate	58 m^3/h	55 m^3/h	41 m^3/h	18 m^3/h
Mass re-suspended	Not detectable	Not detectable	Not detectable	Not detectable
Conc increase	Not detectable	Not detectable	Not detectable	Not detectable
Background conc	25 mg/l	25 mg/l	25 mg/l	25 mg/l

Table A3.12 *Measurements of turbidity around an auger dredger (Bosland, 1991)*

CAUTION SHOULD BE TAKEN IN USING THIS DATA (see Section 3.3.1)

Auger dredger				
Location	Geulhaven, Rotterdam			Ankeveense
Date	June 1991			December 1992
Wormwheel size	4 m long, 1 m diameter			1 m long, 0.4 m diameter
Barge/hopper	VOW56			-
Material type	Consolidated silt			Not measured
In-situ density	1.40 kg/m^3			Not measured
Number of trials	3			1
Measurements	Concentration profiles			Concentration profiles
Water depth	6 m			2.5–3 m
	Trial 1	Trial 2	Trial 3	Trial 1
Trial description	Speed 3.12 m/min 1.5 m/min	Speed 2.94 m/min 1.5 m/min	Speed 2.40 m/min 1.5 m/min	Not recorded
Silt screen	No	No	No	No
Gross production rate	240 m^3/h	240 m^3/h	240 m^3/h	700 m^3/day (in-situ)
Mass re-suspended	Not detectable	Not detectable	Not detectable	Not detectable
Concentration increase	Not detectable	Not detectable	Not detectable	58 mg/l close to bed, otherwise very close to background
Background conc.	18 mg/l	18 mg/l	18 mg/l	20 mg/l

A4 Recent field studies

This appendix describes some of the more recent monitoring programmes that have been undertaken. All of the studies described have made use of acoustic backscatter techniques. Aspects of some of the studies remain commercially confidential.

A4.1 PLUME MEASUREMENT SYSTEM OF THE USACE DREDGING RESEARCH PROGRAMME

In 1988, the US Army Corps of Engineers (USACE) established the Dredging Research Program (DRP) as a seven-year inter-laboratory research effort to develop technologies to reduce the cost of dredging. Two major field experiments have been undertaken by the USACE using acoustic profilers to determine suspended solids concentrations.

These experiments represent the first use of the instrument in this manner and have been carried out as part of the Plumes Measurement System (PLUMES) under development by the DRP Technical Area One.

The first study (USACE, 1991), the Mobile, Alabama, field data collection project, had the following objectives:

- to collect comprehensive data on sediment dynamics for verifying and improving numerical simulation models of the short-term fate (minutes to hours after release) of dredged material placed in open water
- to investigate and refine sediment plume monitoring procedures
- to evaluate acoustic instrumentation for measuring sediment plume dynamics
- to collect field data on coastal bottom boundary layer processes.

USACE (1991) presents the data for meeting the first three objectives and provides a good database for comparison with numerical simulations. It also contains information for conducting similar field exercises. Simultaneous measurement of backscatter intensity from the suspended sediments by two independent and different acoustic systems, made together with water sampling, allowed inter-comparison of the acoustic instruments and provides a first step towards field calibration of the acoustic technique. The recorded plume dynamics include the initial descent, bottom surge, generation of internal waves and evolution of the plume under a variety of conditions.

The second study (USACE, 1992) was undertaken to monitor the movement of dredged material placed with a single-point pipeline discharge. The placement was in an area adjacent to a major oyster seeding ground in the Chesapeake Bay Estuary, an area of environmental concern. The programme of monitoring built on the work undertaken in the study described above. The objectives of the study were to:

- to collect sediment concentration and current data to determine the potential for dredged material to reach the oyster seeding ground
- to continue development of the PLUMES monitoring procedures for dredged plumes.

The field measurements included two days of background monitoring prior to dredging operations, and three days of monitoring during dredging operations.

The background concentration data showed that during the peak ebb and flood phases bottom sediment was re-suspended into the water column. During the dredging acoustic monitoring did not detect dredged material migrating onto the oyster grounds, and water samples showed no alteration of suspended sediment concentration in these areas.

A4.2 CHANNEL DEEPENING TO LONDONDERRY, LOUGH FOYLE, NORTHERN IRELAND

Capital dredging

In 1992, Londonderry Port Harbour Commissioners through the Anthony D Bates Partnership commissioned HR Wallingford to undertake long-term monitoring of suspended solid concentrations at three locations near dredging works associated with deepening of the navigation channel to Londonderry. An upstream site on a disused jetty was selected together with a seaward site on one of the light towers adjacent to the navigation channel. A further site was selected alongside a cooling water inlet for the nearby power station.

Two Phox transmissometers were installed at each of the three sites and a local fisherman was employed to clean and check the installations periodically and report back to HR on their operation every three days. Every three weeks HR personnel undertook routine calibration and downloading of the loggers.

The monitoring was continued throughout the period of dredging and for a further period of one month after completion of the dredging. The long-term monitoring was able to demonstrate that there were no significant increases in suspended solids concentrations in the vicinity of the cooling water intake (HR Wallingford, 1993a). A strong correlation between tidal range and suspended sediment concentration was observed. Analysis of the data collected was unable to isolate any correlation in suspended solid concentration with freshwater flow, wind/wave action or dredging activity.

Suspended sediment concentrations at the three sites varied between 50 mg/l and 700 mg/l on spring tides and between 10 mg/l and 50 mg/l on neap tides. The mean tidal concentration varied between 20 mg/l and 350 mg/l on the flood tide and 20 mg/l and 150 mg/l on the ebb tide over the measurement period.

A further issue associated with the work was the question as to whether sediment re-suspended during the deepening of the approach channel was confined to the channel or whether it spilled out of the channel into the shallow waters either side of the channel. Analysis of the data at the seaward station adjacent to the channel was unable to identify any increases in suspended sediment concentration coincident with passage of the trailing dredger. However, the instrument was deployed so as to sample at ten-minute intervals, so identification of very short-term increases of suspended sediment concentration might not have been possible.

To investigate further the re-suspension from the dredged approach channel, HR Wallingford was commissioned to use a combined suite of instruments to visualise and quantify increases in suspended sediment concentration associated with the different dredging plant being used at the site. The work was undertaken in conjunction with Wimpey Environmental, which supplied and operated the acoustic profiler, and dredging contractor Westminster Dredging, the. During the field measurements, staff from Hydronamic, the environmental engineering section of Boskalis, were on site in Ireland.

The work is described in detail in HR Wallingford (1993a). A summary of the results of the monitoring has been presented in Teal *et al* (1993) and has also been further discussed by Bates (1996).

An RDI 1200 kHz narrow-band acoustic profiler was installed to the port side of Westminster Dredging's vessel *Plover* and used to follow the evolution and fate of sediment plumes re-suspended by various forms of dredging activity being undertaken. Transmissometers, remotely triggered water samples and pumped water samples were used to quantify the suspended sediment concentrations within the plume.

The acoustic profiler was used to identify the location of the plume. For technical reasons, no attempt was made to calibrate the backscatter signal from the acoustic profiler in terms of suspended sediment concentrations. Although successful attempts have been made to quantify suspended sediment concentrations from backscatter signal strength with RDI instruments, none at the time of the Londonderry measurements had demonstrated a calibration in such a dynamic situation in the immediate vicinity of dredging activities.

The dredging plant that was successfully monitored were:

- the trailing dredger *Cornelia*
- the trailing dredger *WD Medway*
- the grab dredger *WD Dredgewell*
- the plough dredger *Norma*.

Maximum concentrations of 3200 mg/l were observed close to the bed behind the *Cornelia*. Near-surface measurements at this stage indicated concentrations of about 500 mg/l. While the *Cornelia* was dredging in the navigation channel, it was possible to detect a plume, apparently emanating from the bed near the draghead. The plume was about 20 m wide and gave a stronger backscatter near the bed. The plume was found to encroach onto the side slope but not to spill up the slope out of the navigation channel into the shallower water adjacent to the channel. After 20 minutes it was no longer possible to detect the passage of the dredger or the plume generated by the dredger with the acoustic profiler or the other water sampling equipment.

Experiments were carried out while monitoring the *WD Medway* to investigate the effects of the passage of the vessel when it was not dredging. It was found that a strong plume was observed in the backscatter record that was confined to about the top 3 m of the water column. This was attributed to aeration. Similar results for the *WD Medway* were determined as for the *Cornelia*, namely that the plumes generated by the dredging decayed to concentrations that were close to background levels over about 20 minutes.

Monitoring was carried out in the vicinity of the *WD Dredgewell*. Measurements about 10 m from the point of dredging demonstrated peak concentrations of 3500 mg/l close to the bed and 250 mg/l at the surface, which was close to background levels. At a distance of 150 m downstream from the dredger it was no longer possible to identify the plume generated by the dredging. It was noted that a complete cycle of operation for the grab took about one minute to complete and it was possible to clearly identify this periodicity at a short distance from the dredger in the record of backscatter and suspended solids concentrations measured with the transmissometer.

A number of acoustic profiles were carried out around the plough *Norma*. It was very difficult to interpret the backscatter from these profiles. This was considered to be due to the turbulence and aeration associated with the plough operation. Water sampling was unable to detect any significant re-suspension in the vicinity of the ploughing. For some of the time *Norma* was working in sandy areas of the channel which, if re-suspended, would be less remain in the water column for significantly less time than silt.

A4.3 AGITATION DREDGING AT SHEERNESS, EASTERN ENGLAND

Capital dredging

Further opportune measurements were made by Wimpey Environmental and HR Wallingford in July 1993, in conjunction with the Port of Sheerness, while the *Agem One* carried out capital dredging at Sheerness (Teal *et al*, 1993). The operation of the *Agem One* is fairly simple: material is sucked up from the bed, passed through a series of disaggregators on deck, aerated and then discharged into the surface waters. The aim is that the aerated mixture will remain as a near-surface plume and be advected away by the tidal currents at the site.

An RDI 600 kHz broadband acoustic profiler was mounted on the side of the survey vessel *MV Medway Surveyor*. The acoustic profiler was configured to collect measurements from 1.8 m below surface to 1 m above seabed at 0.5 m intervals. Signals were transmitted at intervals of 0.09 s and averaged over 4 s.

Measurements were collected during the ebb tides of 13 and 14 July 1993 since dredging operations were limited to the ebb tides only. In addition to the measurements, transmissometer readings and surface water samples were collected during the ebb tide of 14 July 1993. Measurements were collected using a Partech IR40C transmissometer. Water samples were collected using a Cassela Water Sampler.

Acoustic measurements would normally be made as the vessel steamed across the dredged plume at intervals along its length, thereby defining the limits of the plume both laterally and longitudinally. However, in this case, the dredger was positioned alongside the berth, so it was not possible to conduct transects across the plume. Instead, transects were collected longitudinally along the plume and in zig-zags to define its spatial distribution.

Acoustic measurements, transmissometer readings and water samples were collected while the vessel held station within the plume. Instantaneous transmissometer readings were collected at 0.5 m intervals through the depth of the plume. The transmissometer was positioned to coincide as closely as possible with one of the beams of the acoustic profiler. Transmissometer readings were recorded to the nearest 5 s. Profiles were collected immediately downstream of the dredger, 150 m downstream of the dredger and 300 m downstream of the dredger in order to obtain comparison data over a range of conditions.

The acoustic profiler provided good resolution of the plume, but it was not possible to differentiate between backscatter from air bubbles and the sediment re-suspended. It appeared possible to identify the rising air bubbles with distance from the dredger. Initially the solids/water/air mixture descends as a density plume. The solids are then mixed into the water column and advected. Air bubbles however, begin to rise.

A4.4 CONSTRUCTION OF PIPELINE TRENCH, SOUTHERN ENGLAND

Capital dredging

Work was undertaken in conjunction with Boskalis Westminster during mid-1994 while a pipeline trench was being constructed on the south coast of England. The operation involved dredging through clay and chalk using the cutter suction dredger *Orion*. Material was pumped from the *Orion* to the spray pontoon *'s Gravenhage* at a rate of about 400 kg/s. From there it was diffused to the bed to form a temporary storage mound on the seabed some 200 m to the east. After construction of the trench and laying of the outfall, the dredging plant changed positions so that the temporarily stored material was replaced over the outfall pipe in the trench. It is understood that it was not necessary to obtain any additional backfill material for the operation. Most of the temporarily placed material could thus be reused.

The monitoring that was undertaken involved the use of a similar suite of instruments as for the Londonderry monitoring. Measurements were made around both the cutter suction dredger and the spray pontoon while dredging through chalk.

Although the plumes of re-suspended sediment released during the dredging were highly visible because of the chalk content, apart from very close to the operation, the concentrations of re-suspended sediment were observed to be generally less than 30 mg/l compared to a background concentration of 10 mg/l. These figures indicated a flux of material from the cutting operation of less than 0.5 per cent of the production rate. Re-suspension from the spray pontoon during the period of measurement was considered to be about ten times greater than that from the cutter. The detailed results of these measurements remain confidential to Boskalis Westminster and are reported in HR Wallingford (1994a).

This brief description of the study has been provided here to demonstrate the application of the combined water sampling and acoustic technique to other dredging plant.

A4.5 MAINTENANCE DREDGING ON THE RIVER TEES, NORTH-EASTERN ENGLAND

Maintenance dredging

The opportunity arose in mid-1994 for HR Wallingford to undertake a series of controlled experiments during maintenance dredging of the River Tees. The Tees and Hartlepool Port Authority made available its dredger the *Hoertenesse* and a survey vessel. Monitoring, using an acoustic profiler and standard water sampling methods, was undertaken while the dredger was working in both sandy and muddy reaches of the estuary.

Following the field measurements, detailed analysis of the acoustic profiler measurements was undertaken (HR Wallingford, 1995a), to estimate sediment fluxes through the monitored section of the estuary. Following this analysis, a Gaussian diffusion model, which had been used for dispersion assessment studies, was applied to determine whether the model could represent the observed fluxes (Gravelli, 1995). The study clearly showed the ability of the simple Gaussian diffusion model to represent the observed fluxes of material when a variable input source from the moving dredger was used. A source of 20–30 kg/s of mud was applied when the dredger was trailing and overflowing in the muddy reaches of the estuary.

A4.6 AGGREGATE DREDGING OPERATIONS IN THE UK

Aggregate dredging

Numerous assessments of the dispersion of fine sediment re-suspended during marine aggregate production have been undertaken in the UK. The source terms for the dispersion studies come from an analysis of bed material type and discussions with the dredger operator as to the likely operating scenario. In recent years several dredging contractors have measured overflow during the dredging operations. As part of this project, measurements and sampling of overflow and screening discharge were undertaken during dredging from in various licensed areas along the south-east coast of the UK, notably at Owers Bank in the English Channel. They were made from two sizes of trailing suction hopper dredger, during different types of operation. The results are reported in full in MMS (1998). The important feature of these series of measurements is that the results enable an estimation of the quantity of different fractions of material being released during aggregate dredging operations.

In 1995 HR Wallingford, Coastline Surveys and the main UK aggregate dredging contractors carried out a dedicated field survey at Owers Bank using an acoustic profiler and water sampling techniques. This examined suspended solids concentrations in the plumes of sediment re-suspended during dredging as a means of verifying and refining the predictive techniques adopted in earlier environmental assessments of the impact of turbidity generated during dredging. HR Wallingford (1996b) details the study's results. Some of the results of the studies described above have been presented in Hitchcock and Dearnaley (1995). The results presented in the remainder of this section have been taken from that paper. The measurements obtained can be summarised as:

- approximately 150 acoustic profiler transects providing through depth current speed and direction and backscatter data (which can be processed to give additional information regarding suspended sediment concentrations)
- approximately 150 water samples subsequently analysed for total solids content, sand/silt content and particle size distribution
- optical silt monitor output during measurements (c 12 hours of output at 2 s intervals)
- six bed grab samples from one of the vessel tracks
- opportune underwater video showing drag head trails along one of the vessel tracks.

The measurements demonstrated that while it was possible to track a plume for up to 3.5 km from the dredging area (using acoustic profiler monitoring, which can detect very small changes in concentration within the water column) the concentrations within the plume had decayed to background levels of less than 10 mg/l over a distance of less than 500 m. The majority of the material re-suspended was sandy, but the decay of mud and sand concentrations in the plume was found to occur over similar timescales. The study clearly demonstrated a very rapid reduction in suspended sediment concentrations in the immediate vicinity of the dredger.

The field observations were supported by advection-diffusion plume modelling using the GAUSSIAN model. Only by assuming that 20–30 per cent of the material released from the dredger via spillways and reject chutes was initially introduced into the water column could the model reproduce observed concentrations close to the dredging point. The model still over-estimated concentrations at greater distances. These results supported the theory that density and momentum differences between the overflow and water column in the first few minutes of re-suspension and/or aggregation of sediment are the most important factors controlling short-term re-suspension of dredged material.

It should be noted that the acoustic profiler monitoring was carried out at a licence area that contains between two and three times the silt content of other licensed areas.

A4.7 DREDGING ACTIVITIES IN HONG KONG

Capital dredging

Hong Kong is engaged in a major programme of construction that includes reclamation. This has resulted in a need for large quantities of marine sand for fill. It is estimated that about 210 million t has been utilised in this way since 1990 (Whiteside *et al*, 1995). There has been an increasing requirement in Hong Kong to undertake EIAs associated with marine mining. Because of the scale of the operations, this has promoted several studies directed at establishing the losses during the dredging operation and the subsequent advection and dispersion of the plumes of fine sediment so generated. HR has multilayer flow models of Hong Kong waters and employs these as the advective force for examining the fate of plumes of fine sediment re-suspended by dredging.

Specific field measurements have been carried out to examine re-suspension during dredging. These have recently been reported by Whiteside *et al* (1995). This paper presents the 90-minute loading of the 8225 m^3 trailing dredger HAM 310. It was shown that for a total measured load of 11 500 t the total overflow loss was 3000 t. The overflow was calculated as the difference between the inflow and the load. It was estimated that approximately 50 per cent of the overflow was fine particles (<0.063 mm). With these figures it can be seen that approximately 80 per cent of the material dredged was retained and that the average rate of overflow over the 90-minute period was 280 kg/s. If an overflow rate of 7 m^3/s were assumed, the average concentration of fines in the overflow would be 40 kg/m^3, and much of the flow would descend directly to the seabed as a density flow.

Recently, through field measurements using acoustic techniques (Land *et al*, 1994), it has become apparent that the processes occurring during the first few minutes of the plume generation are likely to be responsible for the loss to the bed of a considerable proportion of the fine sediment initially released into the water column (Whiteside *et al*. (1995). HR Wallingford (1995b) has investigated the processes that may be occurring during this initial phase of the development of a plume. The unpublished HR work has been summarised by Whiteside *et al* (1995). They have concluded that, for the vessels operating in Hong Kong with a single sub-surface spillway, the initial momentum of the discharge from the vessel is a significant factor, resulting in much of the sediment descending directly to the seabed. Additionally, it has been postulated that the disaggregation of fine muddy material during the dredging process is not complete and that a further significant proportion of the muddy material released into the water column may be in the form of fine clay balls, or adhered to coarser grains. Thus much of the re-suspended fine material may settle to the bed with settling velocities in excess of that of a natural muddy suspension.

A4.8 MINIPOD MEASUREMENTS AT AREA 107 AND RACE BANK

Minipods are large (2 m in height) self-contained bed frames, which can be deployed onto the seabed with sensors attached, enabling measurements of current speed and turbidity to be taken over periods of several weeks. Between 1994 and 1998 some of these devices were deployed at locations around Area 107, off the UK coast from Skegness, to investigate the dispersion of sediment arising from aggregate dredging. Some 6–7 km to the north-east of Area 107 lies Race Bank, which provides an over-wintering ground for hen crabs. The deployments included a series of four minipods aligned in a series between the dredge site and Race Bank and showed that the turbidity increase arising from the passive plume could be tracked as it progressed from one location to the other. Turbidity increases at Race Bank were found to occur only on spring ebb tides. The tidal excursion on neap tides was too small to transport the plume to Race Bank, and the time taken for plumes to reach Race Bank corresponded to spring tide current speeds. Calibration of the minipod backscatter sensors was undertaken using *in situ* syringe samples and sediment traps and laboratory turbidity tanks. While near-bed concentrations at the dredge site were estimated to be about 700 mg/l (compared with background concentrations of 20–25 mg/l), near-bed concentrations at Race Bank estimated to increase to 50–150 mg/l under certain conditions.

A4.9 MONITORING ASSOCIATED WITH THE ØRESUND LINK, DENMARK AND SWEDEN

Capital dredging

The Øresund Link connecting Copenhagen with Malmo crosses the environmentally sensitive Øresund. The construction of the link involves some major dredging and land reclamation (a total of 7 million m^3 is estimated (Jensen *et al* 1995). The background suspended solids concentration in Øresund is low, generally 1–3 mg/l. A restriction was placed on all dredging operations in the Øresund project, partly because there was limited supply of fine-grained sediment to the area and partly due to the practice of intensive sewage treatment of Sweden and Norway (Jensen *et al*, 1995). There was an overall limit of 5 per cent spill with conventional dredging equipment set 200 m beyond the limits of the dredging area. Additionally daily and weekly restrictions have been placed on maximum spillage rates along the alignment of the link. A spill monitoring programme based on numerical modelling of the sediment spill was been proposed and implemented by the Danish Hydraulic Institute (DHI).

The implementation of the monitoring programme required the use of an acoustic profiler and turbidity monitoring to determine current velocities and suspended sediment concentrations along a number of profiles in order to derive calculate sediment fluxes. At first it was envisaged that backscatter from acoustic profiler measurements could be used to measure suspended sediment concentrations. Weiergang (1995) describes in some detail the analysis method that was been adopted for determining suspended sediment concentrations and fluxes from the acoustic backscatter signals. Although this paper concluded that the uncertainties were insignificant, Jensen *et al* (1995) found that the use of the measured data produced an estimated uncertainty of 25 per cent in the measured sediment flux. The decision was taken by the project management to instead use optical turbidity monitors that were considered to be much more accurate than the acoustic profiler at measuring suspended sediment concentrations.

The project defined a rigorous and consistent approach to measuring spill from dredging activities, with constant monitoring of suspended sediments along transects set at 200 m from dredging enabling the sediment flux and therefore total spillage arising from dredging to be known and compared against the set project limitations all the way through the dredging project (Jensen, 1999). A real-time early warning system for identifying periods when spill would be excessive was implemented, allowing the refined management of dredging to the limitation set by the project. At times this meant dredging had to stop temporarily, especially during the winter months when the currents, which are predominantly driven by barometric pressure systems, were at their highest.

The dredgers used during the project were of the dipper, cutter suction and backhoe type. For each of these dredgers the measured sediment concentrations, fluxes and hydrodynamic conditions are stored on a central database, together with additional data regarding the type of material dredged and the dredging environment. Briefly, the results of the monitoring showed that current speeds at the time of dredging, the depth of dredging and the dredged material had a significant effect on spillage as well as the method of dredging. The cutter suction dredging, mostly of limestone, was found to produce 50–100 per cent more spillage than dipper and backhoe dredging which mostly removed clay till (Lorenz, 1999).

A5 Modelling techniques

This section provides descriptions of the various modelling approaches used for different aspects of dredging plume prediction. Water quality assessment using models to predict the effects of plumes is also discussed. A comprehensive review of techniques is outside the scope of this report, but the existing state of knowledge is described for the purposes of identifying future research needs (see Sections 6 and 7).

A5.1 DREDGING PROCESS MODELLING

As previously noted in Section 4.1 a large quantity of research has been undertaken in the UK and in the USA to characterise the operational effects of dredging on release of fine sediment into the water column. The intention is that these models will enable the prediction of losses for input to plume dispersion prediction studies. These models have utilised the considerable amount of field information that has been established over the last decade (summarised in Appendix 3). However, the usefulness of these models has been hampered by the lack of good quality, consistently measured data (see Section 3.2.1), which has prevented the calibration and validation of these models.

As part of a project for VBKO, Dredging Research (HR Wallingford, 1999b) has created mathematical models that will predict *a priori* the release of sediment caused by grab, cutter suction, bucket ladder, backhoe and trailing suction hopper dredgers. The details of these models are commercially sensitive. The first phase of the project, which reviewed the current state of knowledge and available field data, concluded that latter could not be used for the purposes of calibration and validation of the models. The second phase of the project will include the collection of good quality and consistently measured field data.

Collins (1995) details the formulation of mathematical models for sediment releases made by grab dredgers and cutter suction dredgers. The work is available in the public domain. The basis of the models is the characterisation of the size and strength of the sediment source and the velocities associated with release in terms of dredger operational parameters. This characterisation was achieved by using regression analysis.

A5.2 DYNAMIC PLUME MODELLING

Most of the work undertaken to date has focused on plumes formed during disposal by disposal. Theoretical studies have been undertaken by Koh and Chang (1973) for the USAWES in the USA and by Krishnappen in Canada (HR Wallingford, 1999b).

The Koh-Chang model assumed that the dumped material behaves as a fluid and may thus be an appropriate basis for a model of the dynamic stages of overflow plume development. USAWES developed the approach as a software routine (part of the ADDAMS suite of models – Automated Dredging and Disposal Alternatives Management System) to model disposal operations, but some researchers have found results differ from field data.

The Krishnappen model treats the material as discrete soil particles and is not likely to be appropriate for most overflow operations, with the possible exception of very coarse releases during certain type of aggregate dredging. Laboratory studies leading to empirical models have included work by Ogawa in Japan (HR Wallingford, 1999b). This was also more suited to coarse material and did not include the effects of entrainment and dilution. His model is thus unlikely to be of much use.

Johnson's model (DIFID) (Johnson *et al*, 1993, Johnson and Fong, 1994) predicts the motion of a descending discharge during convective descent, dynamic collapse and passive diffusion. It characterises the material as clouds composed of particles having a specific fall velocity. Clay-silt particles are assumed to be cohesive and the clay-silt lumps are assumed to behave as individual particles with a fall velocity of 0.15 m/s^{-1}. The initial dynamic plume part of the model has been used to form the basis of other dispersion models such as STFATE, which models the dispersion resulting from disposal, and DREDGEMAP (See A5.5).

As part of a study regarding the plumes resulting from aggregate dredging HR Wallingford (1999a) employed a numerical model of the initial phase of plume dispersal developed from USEPA effluent outfall plume models (USEPA, 1985) which are tailored to predicting the average dilution of outfall plumes as they ascend. The rate at which mixing occurs between these two bodies of water was assumed to be proportional to the relative velocity between the plume and the ambient water and the surface area of the plume – a result based on laboratory experiments conducted by Morton et al (1956).

Further research (HR Wallingford, 1999b) led to the development of a prototype dynamic plume model that reproduced the rapid descent of the dynamic plume. It included a criterion involving suspended sediment concentration for deciding the amount of fine material released into the water column. However, as in the previous modelling described above, the model did not include re-suspension caused by the impact of the dynamic plume on the bed.

Applied Science Associates has developed DREDGEMAP, which predicts the transport and dispersion of the sediment through initial dilution and spreading of the sediment in the immediate vicinity of the release. In a second stage it estimates the growth and dilution of the release cloud as it hits either the bottom or a strong density gradient in the water column. In the final stage (passive diffusion) it predicts the transport and dispersion by currents and turbulence. The near field is based on Johnson's DIFID (see Section A5.2). The input includes dredged material density and composition, release location, rate and duration; and the current and hydrographic fields. The model output gives dilution, plume centre, location and width, trapping depths and concentrations.

USAWES has developed STFATE (USAWES, 1995), which again represents three phases of plume dispersion and is based on Johnson's DIFID. The model is freely available to the public.

A5.3 MUD FLOW/CONCENTRATED SUSPENSION MODELLING

Applied predictive models do not yet successfully reproduce the modelling of concentrated suspensions, with the related processes of stratification, entrainment of clear water into the suspension, collapse of turbulence resulting in fluid mud, flocculation and re-suspension. However, some research models, mostly of the 1DV type, reproduce some or all of these aspects of behaviour. The complexity of this type of modelling can be considerable and there is a need for the results gleaned from the research-type modelling and from laboratory work to be incorporated into applied models to enhance their predictive ability. This is the focus of much research, including most notably the European COSINUS project. COSINUS has endeavoured to undertake laboratory and fieldwork and computational modelling to increase the understanding of all aspects of cohesive sediments and in particular to parameterise the results of this work in applied computational models. The work is due to be presented as part of the international INTERCOH conference in September 2000.

The modelling of fluid mud, which for dredging plume situations can occur as a result of the gradual loss of momentum of the concentrated suspension resulting from dynamic plume impact, is more developed than that of the concentrated suspension. Although knowledge of concentrated suspensions and all of the corresponding processes is limited, the behaviour of fluid mud can be characterised reasonably simply as being subject to the processes of advection through pressure gradients, entrainment, re-suspension and dewatering. Field data describing each of these processes exists and can be incorporated into fluid mud models, which can be helpful in predicting, for instance, sedimentation in approach channels. However, the behaviour of fluid mud can vary enormously with different types of cohesive sediment and the most reliable method of prediction of fluid mud is still considered to be observation through field monitoring (HR Wallingford 1999b). The results of the COSINUS project are also expected to benefit the representation of the key processes in these fluid mud models.

A5.4 PASSIVE PLUME MODELLING

Importance of source terms

The most important feature of any passive plume model is the source term used to represent the release of fine sediment into the water column and the choice of this parameter outweighs any of the considerations outlined below. The type of dredging operation together with the type of sediment material can effect the rate of release of sediment by over an order of magnitude (see Table 3.2, MMS, 1998). The effect of water depth, initial density of sediment mixture and ambient currents on the behaviour of the dynamic plume (resulting from trailing suction dredging) can also affect the release rate of sediment into the water column for dispersion as a passive plume by an order of magnitude (HR Wallingford 1999a, 1999b).

Random walk/advection diffusion techniques in numerical models

There are two approaches to modelling the diffusion of sediment from passive plumes. Both generally involve the post-processing of previously modelled flow fields. The first is the advection-diffusion approach, where the equation for movement and diffusion of material is solved numerically. The second, the random walk approach, consists of the tracking of a large number of particles representing the released sediment. The advection is provided by the flow field, while the diffusion is reproduced by the random additional 3D movement of particles. For sufficiently large numbers of particles the random movement is equivalent to the diffusion of a plume.

There are some practical aspects to the choice of either approach. Source terms may be represented differently by different approaches and therefore a particular method may suit a particle study. The random walk method needs a large number of particles to be tracked separately throughout the simulation of plume dispersion and in some circumstances the number of particles needed for the plume to be well represented can be costly in terms of run time.

There are also more fundamental issues associated with each approach. The advection-diffusion method can suffer from numerical diffusion and cannot resolve plumes finer than the (often relatively coarse) hydrodynamic model grid. The resolution of plumes in the random walk approach, however, can be very high if a fine output grid is chosen.

Hydrodynamic input to passive plume models

Finite-difference (rectangular, orthogonal or curvilinear) and finite element models all have their benefits and disadvantages, but any distinction in the accuracy of these models when applied appropriately is small and unimportant.

Other types of computational modelling

Besides particle tracking, as in the random walk method, or numerical solving of advection-diffusion equations it is possible to use analytical methods to reproduce the movement of dredging plumes. One such method is the GAUSSIAN model, explained briefly below.

Representation of dredging activities

The more closely the representation of the release of sediment to reality, the more representative the predicted plume behaviour will be. A discussion of the importance of source terms has been made above. Other considerations such as the position in the water column at which release occurs is really a feature of the dynamic plume model rather than a passive plume model. One improvement that can easily be made to the representation of dredging in the passive plume model is to represent the movement of the dredger as a source of sediment release. The resulting increases in suspended sediment concentration will be different for dredgers moving with or orthogonal to the flow and moving with or against the current flow.

Box A5.1 *The GAUSSIAN model*

> For spatially uniform water depth, and currents the solution for an instantaneous release of a slug of material into the water is as follows:
>
> $$C(x,t) = \frac{M}{4\pi h t \sqrt{D_x D_y}} \exp\left(-\frac{(x-\xi)^2}{4D_x t} - \frac{y^2}{4D_y t} - \frac{W_s}{h}t\right)$$
>
> where
> C is the is the concentration increase above background
> M is the mass release rate
> t is the time after release
> x and y are the coordinates along and perpendicular to the direction of flow
> D_x and D_y are the diffusion coefficients in the x and y directions
> W_s is the settling velocity
> h is the water depth, and
>
> $$\xi = \int_0^t u\,dt$$
>
> The method of Carslaw and Jaeger (1959) can be used to solve the problem for time-varying release. This comprises the addition of several such solutions for placement at small discrete time intervals resulting in a computational solution, C′, for the required release pattern.
>
> $$C'(x,t) \approx \Sigma_{t_1}^{t_2} f(C, x', t_1, t_2)\delta t$$
>
> where $f(C, x', t_1, t_2) = C(x - x'(t), t)$ $t_1 < t < t_2$
> $\qquad\qquad\qquad\quad\;\; = 0$ otherwise
>
> and x'(t) describes the position of the dredger at time t.
>
> This method assumes the initial uniform distribution of released material through the water column of the sediment and models the settling of material on the bed as a steady stream of material under conditions where shear stress is below the threshold for deposition. The method does not allow for the re-suspension of sediment from the bed as slack water ends and current speeds pick up. The deposition predicted by the method is thus an upper limit and the plume concentrations are an underestimate.

A5.5 WATER QUALITY MODELS

Before describing the existing state of water quality models associated with the prediction of impacts arising from dredging plumes it is useful first to consider the context of the water quality problems that can arise.

Predictive modelling of the changes in water quality due to dredging plumes is generally concerned with the fate of toxic contaminants that can exist in sediment dredged from harbours and ports. Impacts associated with DO depletion or nutrients are not normally significant because of the mixing effect of currents and waves. Exceptions to this can occur in sheltered harbours with little or no current flow.

Contaminants generally recognised to be of concern include inorganic metals such as such as Hg, Pb and Zn and organic compounds such as TBT, pesticides and PAHs. The behaviour of contaminants in dredging plumes is a function of the movement of the plume but also the following:

- adsorption/desorption
- precipitation/dissolution
- complexation/disassociation
- oxidation/reduction.

Parameters called partition coefficients, established from laboratory analysis, are used to model these processes and literature gives a range of values for different contaminants, types of sediment, suspended sediment concentrations, temperatures, salinities, pH and organic matter content. The process of dredging will also have an effect on contaminant behaviour. During dredging, buried anoxic substances become exposed and change their physico-chemical properties, including their adsorption onto sediment. This means that contaminant behaviour is complex and highly site-specific. However, there is no chemical model or site-specific data that will provide information about the correct partition coefficients to use. The values of partition coefficients given in the literature can vary by orders of magnitude. This means that for any given situation there is considerable uncertainty in the description of the contaminant behaviour.

Water quality modelling of the impacts of dredging plumes at present takes a pragmatic approach to this problem. Because of the nature of the uncertainties in the description of contaminant behaviour, full dynamic modelling of the interaction of contaminant with water and sediment is usually inappropriate. Instead the approach is to use the results of plume dispersion models to predict the suspended sediment concentrations, which are then multiplied by the adsorbed concentration of contaminant in the sediment, itself derived from partition coefficients available in the literature. A similar process can be undertaken for desorbed contaminants.

HR Wallingford has carried out water quality assessments of impacts from plumes in Hong Kong (eg ERM 1996, HR Wallingford, 1997) using this approach. In the USA, the Waterways Experiment Station (USAWES) has formulated dredging and disposal dispersion models, DREDGE and STFATE, available to the public, which include short-term water quality calculations based on this approach (USAWES, 1995).

A6 A comparison of predictive approaches of sediment plume dispersion

A6.1　INTRODUCTION

There are three common methods for considering the dispersion of sediment re-suspended from dredging, each of which involves a different level of analysis:

- a desk analysis involving tidal excursion lengths and residual currents to identify the areas likely to be affected by the plume
- use of a basic advection/diffusion model to establish an initial estimate of suspended sediment concentration increases and potential deposition
- full 2D/3D process modelling of sediment transport.

Although the choice of approach depends on the nature of the study, these approaches are not mutually exclusive and best practice usually entails an initial appraisal followed by a more in-depth study. The type of results obtained for each of these methods on different dredging operations is considered below. Although these tools differ as to the level of analysis, the quality of the results determinable by each of the approaches is hugely dependent on the quality of knowledge of the initial conditions, in particular, the mass and rates of input of sediment initially introduced into the water column. However, this shortcoming does not affect the suitability of the process modelling approach, but rather demonstrates the need for more realistic initial conditions for input into these modelling tools. As understanding of the initial conditions of dredging plumes increases, the reliability of the modelling results will correspondingly increase. Best practice therefore requires acknowledgement of the uncertainty of input rates to the passive stage of the dispersion and the use of a range of input rates to demonstrate the consequences of this uncertainty.

A6.2　DESK ANALYSIS

Desk analysis of dispersion from aggregate dredging operations involves an initial appraisal of the plume dispersal resulting from dredging. In such a study one might consider:

- type of sediment being dredged
- system of dredging
- likely proportion of sediment that might be lost to the water column
- magnitude of background suspended sediment concentrations
- likely settling velocity of the released material
- speed and direction of currents at the point of dredging
- tidal excursion (based on the measured speeds)
- residual movement of the plume (based on the measured speeds).

Mechanisms not considered include:

- dispersion of the sediment plume
- spatial (vertical and horizontal) variation in velocity fields
- the ability of currents to prevent settling and to re-suspend sediment
- vertical turbulence (movement of sediment up and down the water column)
- variation in concentration through the vertical profile
- initial density-induced rapid downward movement of the plume.

Desk analysis enables identification of the likely areas that will be affected by the plume, an initial estimate of the amount of sediment that may be lost into the water column and an upper limit for the deposition of sediment away from the release point. Much of the work undertaken during desk analysis constitutes a necessary part of the preparation for computational modelling studies.

A6.3 ANALYTICAL ADVECTION/DIFFUSION MODELLING

The advection/diffusion modelling method provides an economical means of predicting the dispersion and settling of a plume of suspended cohesive sediment in a large uniform area. There are a number of simplifying assumptions inherent in such methods, but they provide a more accurate numerical estimate of the increases in concentration and of an upper limit for deposition than the desk analysis approach.

One example of this type of approach is the GAUSSIAN model (see Box A5.1). GAUSSIAN represents the processes of advection by currents, settling of the sediment through the water column and the diffusion of suspended sediment due to the natural turbulence in the flow. The flow within the water area under investigation is assumed to be uniform and uni-directional along a single axis direction. The depth in the area of interest is also assumed to be uniform. Diffusion along and perpendicular to the direction of flow are input parameters, with the former significantly greater than the latter.

The method assumes the initial uniform distribution of released material through the water column of the sediment and models the settling of sediment on the bed as a steady stream of material under conditions where shear stress is below the threshold for deposition. The method does not allow for the re-suspension of sediment from the bed as slack water ends and current speeds pick up. The deposition predicted by the method is thus an upper limit.

Mechanisms not included:

- spatial (vertical and horizontal) variation in velocity fields (such variation will increase dispersion)
- vertical turbulence (movement of sediment up and down the water column)
- variation in concentration through the vertical profile
- re-erosion of sediment from bed
- initial density-induced rapid downward movement of the plume.

A6.4 PROCESS MODELLING – SEDPLUME

2D and 3D process modelling enables a more realistic estimate of the increases in suspended sediment concentration and of deposition than the advection/diffusion model described above. In particular the inclusion of spatial variation in current velocity, and the processes of turbulent diffusion and re-suspension of sediment from the bed, can greatly improve the description of plume movement. However, the accuracy of process modelling is hampered by the poor state of knowledge of the initial conditions of the released sediment, in particular in terms of the amount of sediment lost during the dredging process that is initially released into the water column to be advected and dispersed. This is the subject of ongoing research by a team being lead by HR Wallingford for Dutch dredging contractor VBKO.

One example of process modelling is the HR SEDPLUME model. Briefly, the model post-processes the hydrodynamic output from an unstructured finite element flow model (TELEMAC). It is a random walk model, and uses the assumption of a logarithmic velocity profile through the water column, or if required, 3D hydrodynamic output, to track the three-dimensional movement of sediment particles. Dispersal in the direction of flow is provided by the shear action of differential speeds through the water column while turbulent dispersion is modelled using a random walk technique. The deposition and re-suspension of particles are modelled by establishing critical shear stresses for erosion and deposition. Re-suspension occurs when the bed shear stress exceeds the critical shear stress for erosion, while deposition occurs when the bed shear stress falls below the critical shear stress for deposition.

The SEDPLUME model does not account for the initial rapid movement of the released plume due to its initial momentum and greater density than the underlying fluid. The model uses an initial condition whereby the sediment is distributed in a gaussian fashion over a specified radius, uniformly distributed throughout the water column or at a specified height above the bed. Where more information about the likely distribution of sediment just after release is known, SEDPLUME will give a more accurate representation of the initial movement of the sediment plume than GAUSSIAN, although it is more common that little is known about the initial circumstances of sediment release.

Assumptions:

- logarithmic velocity profile (when using 2D hydrodynamics)
- depth-averaged suspended sediment concentrations (when using 2D hydrodynamics).

Mechanisms not included:

- initial momentum or density-induced rapid downward movement of the plume.

In the remining sections these predictive methods are compared with respect to different dredging conditions.

A6.5 METHODOLOGY

The three methods – the desk analysis approach, with GAUSSIAN and SEDPLUME representing the advection/diffusion and process-based approaches – were compared using two dredging scenarios (HR Wallingford 1999a). The release is simulated from a position on the south-east coast of the UK where flow conditions are reasonably unidirectional. In this case, release is continual over five tides. As shown in Table A6.5, the following arbitrary, but representative, parameters were used:

Table A6.1   *Model parameters*

Parameter	Value
Diffusion coefficient	$D = 0.3$ m^2/s
Settling velocity	$W_s = 1 \times 10^{-3}$ m/s
Loss rate of fine sediment	$\partial m/\partial t = 10$ kg/s
Critical shear stress for re-erosion	$\tau_e = 0.5$ N/m^2
Critical shear stress for deposition	$\tau_d = 0.1$ N/m^2
Erosion rate (SEDPLUME)	$M = 0.0005$ ms
Dry density of settled sediment	$\rho_d = 500$ kg/m^3

A6.6 RESULTS FOR UNIFORM FLOW CONDITIONS – OFFSHORE FROM HARWICH HARBOUR

The movement of sediment in approximately uniform flow conditions was represented by release at a position some 15 km south-east of the mouths of the Stour and Orwell estuaries. The release was modelled as continual over five (repeating) mean spring tides, starting at HW, and the resultant movement of the plume was modelled using the GAUSSIAN and SEDPLUME models.

The peak mean spring current (depth-averaged) speeds at the (hypothetical) release point vary from 1.0 m/s on the flood tide to 1.2 m/s on the ebb tide with the flood and ebb tidal excursion lengths being 15 km and 17.5 km respectively. The depth of water at the site was 14 m. After 0.5 tides the plumes extended some 17 km (GAUSSIAN) and 14.5 km (SEDPLUME) on the initial ebb tide. The results were similar except for the different distance travelled by the plumes, and at the north extremity of the plume where the SEDPLUME model predicts concentrations above background in excess of 5 mg/l while the GAUSSIAN model only predicts concentrations just above 2 mg/l. The corresponding deposition footprints were similar except that the GAUSSIAN result shows greater deposition towards the release point and less deposition towards the northern extremity of the plume. The difference between the GAUSSIAN and SEDPLUME results is principally caused by deposition occurring during the initial slack time of the model run. This is re-suspended in the SEDPLUME model but is not re-suspended in the GAUSSIAN model. Note also that SEDPLUME predicts that all of the sediment deposited is re-suspended on the following tide, unlike the GAUSSIAN model, which assumes permanent deposition.

After 1.0 tide the results of the two models matched less well than previously. The GAUSSIAN plume concentration fell below 1 mg/l above background some 6 km south-west of the release point. The SEDPLUME plume extended for 15 km south-west of the release point, with concentrations above background in excess of 5 mg/l. Again, the reason for this is the re-suspension of deposited sediment in SEDPLUME, which is not re-suspended in GAUSSIAN. The effect was larger than after 0.5 tide because sediment has been able to deposit during the LW slack as well as the initial HW slack. The GAUSSIAN model assumes a steady fall of sediment at a predetermined rate as long as speeds are below a preset threshold. The model does not allow for re-suspension of this sediment. This type of situation represents quiescent conditions where current speeds are too low to re-suspend deposited sediment, which is not the case in the example quoted above, or alternatively, represents locally quiescent conditions where bed geometry provides protection from re-suspension.

The GAUSSIAN model could therefore be seen as a worst-case scenario that predicts the maximum deposition possible at any point. However, its prediction of the deposition footprint will be unrealistically high as a whole, unless current speeds at the point of interest are very low. The SEDPLUME model is able to represent the effect of re-suspension and therefore the prediction of the deposition footprint is more accurate than that of GAUSSIAN. However, model and survey resolution of bathymetry means that locally sheltered areas are unlikely to be represented within the model. A worst-case scenario is therefore worked out in a similar manner by assuming that all deposited sediment remains settled.

The results show that over the short term, ie the first few hours of plume generation up to the first slack water, GAUSSIAN can give a reasonable representation of plume dispersion in uniform flow conditions. It is usually the case that this period is of primary interest in plume dispersion studies. However, the longer-term dispersion of the plume is a combination of the advection of sediment with the current, the diffusion of sediment caused by turbulence and differential current speeds through the water column, and the loss of sediment deposited on the bed. Essentially the more deposition predicted by a method, the smaller the magnitude of longer-term suspended sediment concentrations above background levels. GAUSSIAN therefore does not give an accurate or worst case for suspended sediment concentration increases in the long term, unlike SEDPLUME, which is a more useful tool for longer-term dispersion.

A6.7 EFFECT OF LAGRANGIAN RESIDUALS

The longer-term dispersion of sediment plumes, even in apparently unidirectional circumstances, can lead to inaccuracies in the modelling of plume advection in simpler, more idealised models. The example used above shows that for short-term dispersion it is the treatment of deposition and re-suspension that governs the accuracy of prediction, with the advection of the plume apparently well represented in both models. However, over a longer period, the lagrangian nature of suspended sediment transport becomes more important. The differences in advection between the models in the comparisons made above were small. As the dispersion continues, however, the small differences arising from the non-uniformity of the current patterns has a significant effect on the movement of the plume. After 4.5 tides the the plume as modelled by SEDPLUME affects areas as much as 5 km to the east of the GAUSSIAN plume. This phenomenon can be an important consideration in an environmental impact assessment.

A6.8 NON-UNIDIRECTIONAL FLOW

The comparisons made above discuss the how the GAUSSIAN method, which is based on uniform flow conditions and proscribes re-suspension, results in much higher deposition near the release point, and consequently lower concentrations away from the release point. Furthermore, the impact of mild non-uniformity of flow conditions over long dispersion times has been demonstrated. The effect of significantly non-uniform flow conditions exacerbates this trend in GAUSSIAN results, compromising the accuracy of GAUSSIAN predictions except in areas local to the release point.

A6.9 OTHER CONSIDERATIONS

The dispersion of sediment can occur in different ways, of which the most common are small-scale temporal variation of currents (turbulent dispersion) and spatial variation in currents. The latter occurs both horizontally and vertically. The variation of current speed through the water column is approximately logarithmic, which results in dispersion as flow near the surface flow is faster than flow near the bed. This type of dispersion is referred to as shear dispersion and is a much larger effect than turbulent dispersion. This type of dispersion must be approximated by a diffusion coefficient in GAUSSIAN, whereas SEDPLUME can reproduce the shear effect.

Further non-uniformity in flow conditions can occur as a result of salinity-induced density gradients in an estuary, or as a result of wind action, which produce opposing residual currents near the bed and near the surface of the water column. Sediment near the bed will have a residual tendency to travel in one direction until turbulence carries this sediment into the upper water column, whereupon the sediment will have a residual tendency to travel in the other direction. The overall residual tendency of a plume is thus dependent on the proportion of sediment in the upper and lower parts of the water column. This type of information can only be deduced if flow conditions are allowed to vary through the vertical profile, and for accurate plume modelling under these conditions 3D hydrodynamic input is required.

A6.10 CONCLUSIONS OF COMPARISONS OF DIFFERENT APPROACHES TO PLUME DISPERSION PREDICTION

1. An initial scoping exercise must be carried out as part of any environmental assessment procedure. A desk assessment of the potential initial pattern of dispersion should be incorporated in this exercise. The scoping exercise should identify whether further modelling (GAUSSIAN or SEDPLUME) is required.

2. For spatially uniform flow conditions and where there are no flow effects resulting from the geometry of the seabed, GAUSSIAN provides an economic and computationally efficient method of calculating upper limits for deposition.

3. The accuracy of the GAUSSIAN prediction improves with shorter simulations and as flow conditions approach those where deposition is continuous.

4. The accuracy of the GAUSSIAN prediction deteriorates as simulation time increases.

5. Where flow conditions are not spatially uniform, where dispersion over long time periods is required, where geometry interferes with flow patterns, where an accurate prediction of suspended sediment concentrations is required, or where re-suspension of sediment is likely to be significant, the GAUSSIAN method cannot describe plume dispersion. In this case process modelling (such as SEDPLUME) is a more suitable option.